pH Control

An Independent Learning Module from the Instrument Society of America

pH CONTROL

By Gregory K. McMillan

INSTRUMENT SOCIETY OF AMERICA

Printed in the United States of America

INSTRUMENT SOCIETY OF AMERICA
67 Alexander Drive
P.O. Box 12277
Research Triangle Park, NC 27709

Library of Congress Cataloging in Publication Data

McMillan, Gregory K., 1946–
pH control.

(An Independent learning module from the Instrument Society of America)
Bibliography: p.
Includes index.
1. Hydrogen-ion concentration. 2. Chemical process control. I. Title. II. Series: Independent learning module.
QD561.M46 1985 541.3′728 84-29976
ISBN 0-87664-725-5

Editorial development and book design by Monarch International, Inc.
Under the editorial direction of Paul W. Murrill, Ph.D.

Production by Publishers Creative Services Inc.

Table of Contents

PREFACE

ISA's Independent Learning Modules

This is an Independent Learning Module on pH Control; it is part of the ISA Series of Modules on Unit Process and Unit Operations Control.

Comments about This Volume

This ILM on pH Control is based on pH system analysis techniques previously documented by V. L. Trevathan and pH signal characterization principles recently documented by D. M. Gray that were extended and applied to numerous new and old pH installations. The experience gained indicates that success depends not only upon proper assessment of the difficulty of the loop and the selection of the appropriate control strategy but also upon the ability to recognize and avoid many potential pitfalls during the specification and installation of the control valve, reagent piping, mixing equipment, and electrodes.

The purpose of this ILM is to instruct the reader on the use of graphical and algebraic techniques for system analysis and signal characterization and to provide extensive knowledge from field experience to ensure successful system design and implementation.

Unit 1: Introduction

UNIT 1

Introduction

Welcome to ISA's Independent Learning Module on pH Control. The first unit provides an overview of the distinguishing characteristics of the pH control problem and the resulting system requirements.

Learning Objectives—When you have completed this unit you should:

A. Appreciate the implications of the extreme rangeability of the pH process variable.

B. Appreciate the implications of the extreme nonlinearity of the pH process variable.

C. Understand the overall methodology for designing and commissioning successful pH control systems.

1-1. Definition of pH

The process variable pH is the negative logarithm of the hydrogen ion activity. For the purposes of this unit, activity can be considered as equivalent to concentration. The units of concentration and the relationship between activity and concentration will be explained in sufficient detail in Unit 2. The small p designates the mathematical relationship between the ion and the variable as a power function; the H designates the ion as hydrogen. Hydrogen ions exist in all streams that contain water or an acid, and can be visualized as a single, positive-charged hydrogen nucleus, which is a proton (H^+). In water streams this hydrogen ion is thought to be actually bound to a water molecule to form a hydronium ion (H_3O^+). At the interface of the pH measurement glass electrode, the hydrogen ion is liberated to join up with a water molecule in the glass (dried out glass electrodes will not respond to pH). This response mechanism and its implications on the performance of electrodes will be explained in Unit 4 to facilitate the design, installation, and troubleshooting of pH measurement systems.

Equation (1-1a) mathematically describes the definition of pH. If both sides of this equation are multiplied by -1 and the definition of an antilogarithm used, Eq. (1-1b) results. Equation (1-1b) shows that positive values of pH correspond to a concentration less than one and that negative values of pH correspond to a concentration greater than one. Concentrated sulfuric acid has a pH of −10 and concentrated sodium hydroxide has a pH of +19. Why then do pH meter scales have a maximum range of 0 to 14 pH? First, the setpoints of feedback pH loops are well within the 0-to-14 range since biological processes and equipment cannot tolerate such high concentrations. Second, even though some feedforward pH measurements would fall outside this range, the pH electrode accuracy rapidly deteriorates at the extremeties of this range. Table 1-1 lists the hydrogen ion concentrations per Eq. (1-1b) for a scale of 0-to-14 pH. Note that the hydrogen ion concentration decreases by a factor of 10 for each unit increase in pH.

$$pH = -\log C_H \tag{1-1a}$$

$$C_H = 10^{-pH} \tag{1-1b}$$

where:
pH = pH
C_H = hydrogen ion concentration (gm-moles/liter)

pH	Hydrogen Ion	Hydroxyl Ion
0	1.0	0.00000000000001
1	0.1	0.0000000000001
2	0.01	0.000000000001
3	0.001	0.00000000001
4	0.0001	0.0000000001
5	0.00001	0.000000001
6	0.000001	0.00000001
7	0.0000001	0.0000001
8	0.00000001	0.000001
9	0.000000001	0.00001
10	0.0000000001	0.0001
11	0.00000000001	0.001
12	0.000000000001	0.01
13	0.0000000000001	0.1
14	0.00000000000001	1.0

Table 1-1. The hydrogen ion concentration decreases and the hydroxyl ion concentration increases by a factor of 10 for each unit increase in pH.

Equation (1-2) shows that the product of the hydrogen and hydroxyl ion concentration is a constant for water at a given temperature. The hydroxyl ion consists of a hydrogen ion bound to a oxygen ion with a net single negative charge. Thus, the charge of the hydroxyl ion is equal in magnitude and opposite in sign to the hydrogen ion. At 7 pH, the hydrogen ion concentration equals the hydroxyl ion concentration so that the net charge of the water is zero. The 7 pH point is termed the neutral point. Table 1-1 shows that the hydroxyl ion concentration increases by a factor of 10 for each unit increase in pH.

$$C_H * C_{OH} = 10^{-14} \text{ (at 25°C)} \quad (1\text{-}2)$$

where:
C_H = hydrogen ion concentration (gm-moles/liter)
C_{OH} = hydroxyl ion concentration (gm-moles/liter)

1-2. Difficulty of pH Control

Figure 1-1 shows that a typical pH control system consists of a set of pH electrodes, a pH transmitter, a feedback controller, a control valve, and a piece of mixing equipment. The input stream whose pH is to be adjusted is called the influent, the acid or base used to do the pH adjustment is called the reagent, and the output stream whose pH was adjusted is called the effluent. The control strategy appears to be relatively simple compared to that for many unit operations. Why then is pH control so difficult?

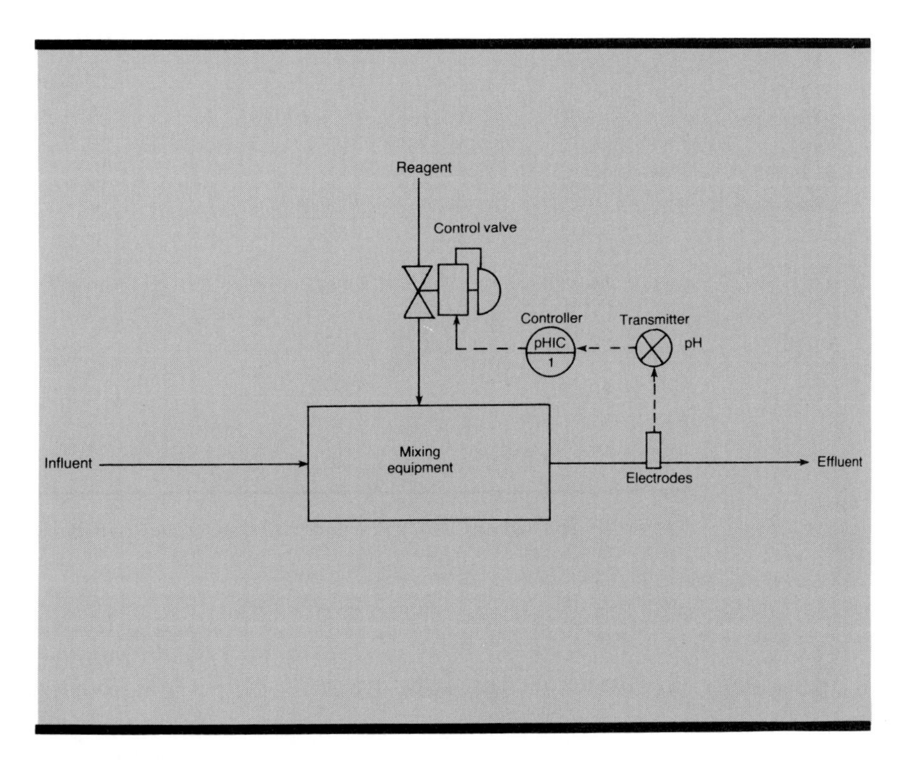

Fig. 1-1. A typical pH control system is simple in strategy but each component requires special attention for the system to work.

Table 1-1 illustrates the heart of the problem. The standard scale range of 0 to 14 pH corresponds to a concentration measurement range of ten to the zero power (1.0) to ten to the minus fourteenth power (0.00000000000001). No other type of commonly used measurement covers such a tremendous range. Also, the pH electrode can respond to changes as small as 0.001 pH, which means the pH measurement can track changes of 0.0000000005 in hydrogen ion concentration at 7 pH. No other commonly used measurement has such tremendous sensitivity. As with most things in life, you don't get something for nothing!

The rangeability and sensitivity capabilities create associated control system design problems that can seem insurmountable. It is important to realize that these problems are due to attempting a level of performance in the pH process in terms of concentration control that goes well beyond the norm. For a strong acid and strong base control system, the reagent flow required is proportional to the difference in hydrogen ion concentration between the influent and the setpoint pH. The reagent control valve must then have a rangeability greater than 10,000,000:1 for an incoming stream at 0 pH and a setpoint at 7 pH. Since the control valve stroke in terms of hydrogen ion concentration translates into pH errors via the titration curve, the dead band of the same control valve must be less than 0.00005% to control within 1 pH of a 7 pH setpoint. How then is this possible? Such strong acid and base systems are controlled by approaching the setpoint in stages and using successively smaller precision control valves. The multiple stage requirement of pH control can be visualized by comparing it to trying to sink a golf ball in the hole on a green. The distance between the tee and the green represents the rangeability requirement and the size of the hole compared to the distance represents the sensitivity requirement. For the above strong acid and strong base system, the tee would be about a million yards from the green. A hole in one is impossible. Using the same large control valve at each stage is like the joke about the gorilla who drives the green in one stroke but then uses his driver again and hits the ball the same distance when he tries to putt the ball in the hole.

The steady-state gain seen by the pH controller is the final change in measurement signal divided by the change in control valve signal. For the strong acid and base system, a fractional change in acid reagent flow (and thus hydrogen ion concentration) from 1.0 to 0.1 at 0 pH causes a change of one pH while a fractional change in reagent flow from 0.000001 to 0.0000001 at

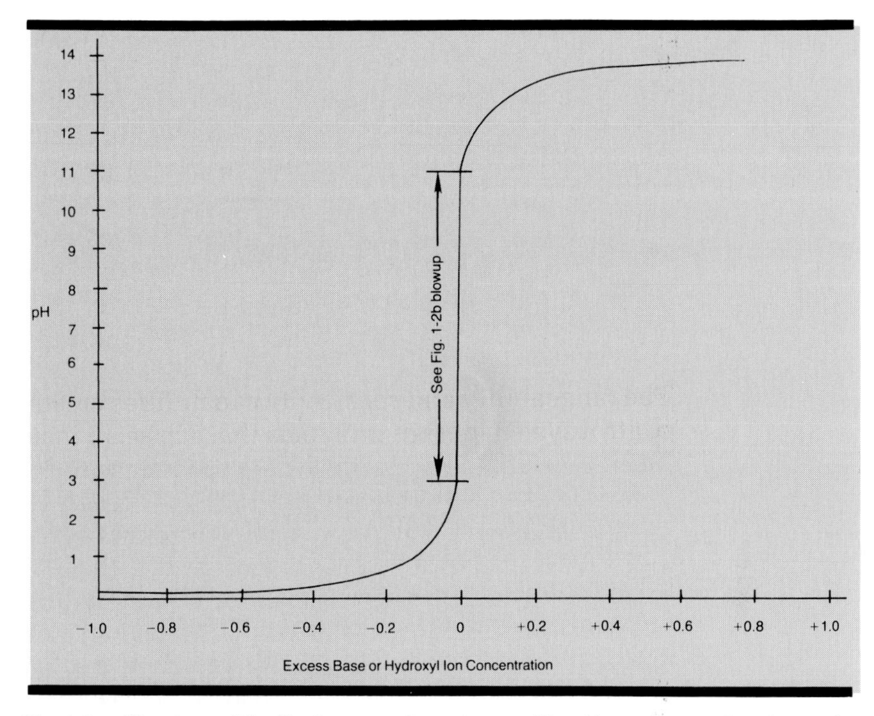

Fig. 1-2a. The slope of the titration curve for a strong acid and base changes by a factor of 10 for each unit deviation from 7 pH.

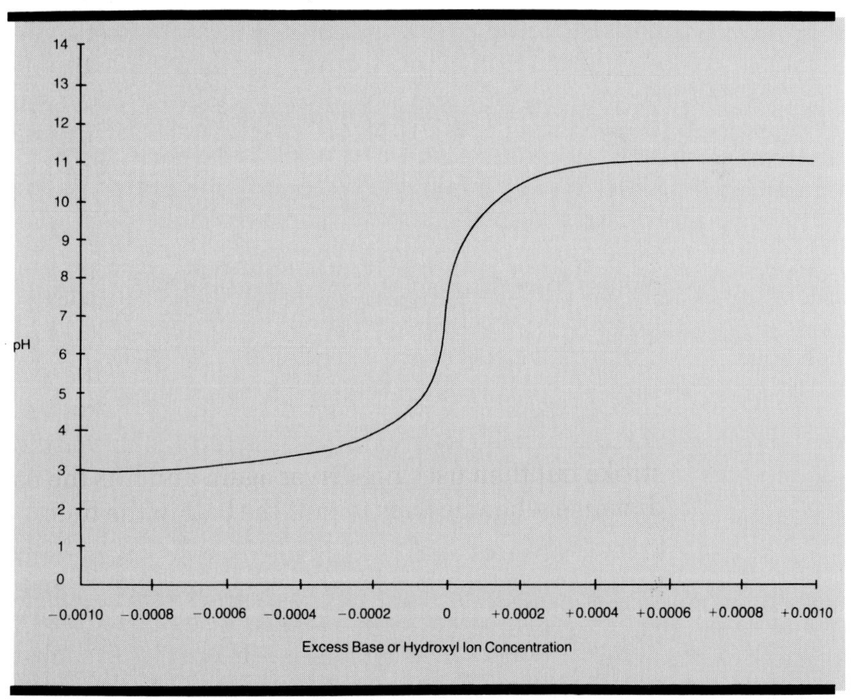

Fig. 1-2b. A blow-up of the vertical neutral region of Fig. 1-2a reveals another S-shaped curve.

6 pH also causes a change of one pH. The pH gain is proportional to 1/0.9 or 1.1 at 0 pH and to 1/0.0000009 or 111111.1 at 6 pH. The pH process gain at a given pH setpoint is best visualized as the slope of the titration curve divided by the influent flow at that pH. The titration curve is a plot with pH for the Y axis and the ratio of acid to base or difference between acid and base concentration, volume, or flow for the abscissa. For a strong acid that provides a single hydrogen ion from each molecule and for a strong base that provides a single hydroxyl ion from each molecule, the difference in acid and base concentration equals the difference between hydrogen and hydroxyl ion concentration. Figure 1-2a shows a titration curve for a strong acid and strong base with the abscissa labeled both with acid and base and hydrogen and hydroxyl ion concentration. The curve appears to be a vertical straight line between 2 and 12 pH. However, a blowup of this region in Fig. 1-2b reveals another S-shaped titration curve. Successive blowups centered around 7 pH would always yield additional S-shaped curves, since the slope is continuously changing by a factor of 10 for each pH unit deviation from 7 pH. This type of graphical deception is a common problem in pH system analysis. The titration curve is symmetrical about the 7 pH point, which is the neutral and equivalence point for this system. Note that the difference in acid and base or hydrogen and hydroxyl concentration is zero and the slope is steepest at this point.

In order for a control loop to respond equally well for all excursions along the 0 to 14 pH scale, the controller gain would have to change in an equal and opposite direction to the pH gain so that the loop gain is constant. The use of multiple tanks to limit the pH excursions to a small region around setpoint reduces the change in gain seen by the controller. The setpoints are incremented between the influent and desired final pH so that the distance in acid or base concentration between the influent and setpoint pH is reduced for each control loop.

In order for the pH control system to work, special attention must be paid to the design and installation of electrodes, transmitter, controller, control valve(s), piping, and mixing equipment—in other words, to each component of the system. A mistake in the design or installation of any component can cause the control system to fail miserably. The process, mechanical, and instrument engineers all must be alert to special system requirements from the very beginning of the project. The most common mistake is to allocate insufficient money for the required number of well-mixed tanks in the project estimate to meet the control objectives of the project premise.

1-3. System Design Overview

Selection of the proper number and type of mixing equipment depends on an assessment of the difficulty of the pH control system. How difficult the control system is depends on the type and concentration of the acids and bases in the influent and reagent and the size and speed of disturbances to the control system. The titration curve is the best tool to evaluate the effects of composition and flow disturbances. pH control systems differ from other control systems in that the disturbances can originate not only from process or pressure changes but also from limitations in flow measurement and control valve stroke precision and fluid mixing due to extreme pH process sensitivity. This is especially true for strong acids and bases. Thus, the number of stages of treatment required also depends on the instrument and equipment capability. The reader is directed to Ref. 1 for a good overview of the importance of the titration curve in determining the pH control system characteristics.

The instrument system design requires information from the titration curve, information on the fluids, and attention to numerous details to avoid potential problems. Information on the effluent pressure, temperature, fouling characteristics, composition, and conductivity is used for designing the electrode installation. Information on the changes in pH gain is used for selecting the controller type. Information on the reagent rangeability and sensitivity requirements and on the fouling characteristics of and pressure fluctuations in the reagent is used for designing the control valve installation.

A decision chart that illustrates the methodology and information requirements for proper equipment and instrument design is presented in Unit 11 as a useful reference that is also capable of being programmed for computer-aided design.

1-4. Applications Overview

Table 1-2 lists some applications and the importance of pH measurement and control. The number of bacteriological and fermentation applications will increase in years to come as biochemicals move from laboratory development to industrial production. Not mentioned in the table is the general use of pH measurement and control for corrosion prevention in equipment and piping and to meet environmental restrictions on plant effluent. For further discussion of the applications in Table 1-2, the reader is directed to Ref. 2.

Application	Processes and Aspects Affected by pH
Bacteriology	Microorganism growth and metabolism
Baking	Dough volume, texture, and color
Brewing	Yield of extract and sugar during mashing
Canning	Time and temperature for sterilization
Chemicals	Impurities and crystallization of salts
Cleaners	Effectiveness of removing paint and varnish
Dyes	Yield and uniformity through intermediates
Electroplating	Nickle deposit hardness and brightness
Fermentation	Fermentation time and alien organism growth
Gelatin	Water absorption, solubility, and clarity
Pharmaceuticals	Effectiveness, stability, and body reaction
Pigments	Uniformity of composition in precipitation
Pulp and Paper	Sizing, loading, coating, and dyeing
Sewage	Digestion time, odors, and foaming
Sugar	Inversion of sugar and destruction of glucose
Textiles	Efficiency of most wet processes
Water treatment	Coagulation and softening processes

Table 1-2. pH is important in many types of industrial production.

Exercises

1-1 *For the strong acid and base system, what is the order of magnitude decrease in acid flow required to go from 1 to 6 pH?*

1-2. *For the strong acid and base system, why would the control error caused by reagent control valve hysteresis be less at 1 than 6 pH?*

1-3. *For the strong acid and base system, what is the order of magnitude change in pH process gain if the pH increases from 1 to 6 pH?*

1-4. *For the strong acid and base system, at what pH is the pH process gain the largest?*

References

[1]Trevathan, V. L., *Characteristics of pH Control*, AIChE Workshop on Industrial Process Control, Tampa, 1974.
[2]*Modern pH and Chlorine Control*, Taylor Chemicals Company, 1945, pp. 57-80.

Unit 2: pH Chemistry

UNIT 2

pH Chemistry

This unit describes how to quantify the effect on pH of various types of acids and bases in solution.

Learning Objectives—When you have completed this unit you should:

A. Be able to convert from most types of concentration units to units for the abscissa of the titration curve that are useful for control system analysis and design.

B. Understand the limits of pH as a concentration measurement.

C. Become familiar with the terms and equations that affect the shape of the titration curve and are used in distributed system controllers to compensate for the pH nonlinearity.

2-1. Ion Concentration

As in most scientific and engineering fields there are several different types of units in use. The units most frequently used to quantify the concentration of acids and bases in solution are molar, molal, weight percent, and normality. Since the material balance and specifications for control systems use reagent and influent flows defined in either pounds per hour or gallons per minute, it is desirable to be able to convert from the concentration units to a variable that uses the flow units. This variable is the ratio of reagent to influent flow and will be used in the remainder of the text for the abscissa of the titration curve for pH system analysis.

Molar units are predominately used by chemists making laboratory measurements because the calculation is based on a beaker volume. Molar concentration is the number of gm-moles per liter of solution. The number of moles is calculated by taking the weight of the pure acid or base and dividing it by its molecular weight expressed in the same weight units. The weight units used should be denoted with a dash in front of the word "moles" (i.e., lb-moles and gm-moles). Equations (2-1a) and (2-1b) show how to calculate gm-moles and then the concentration in molar units. It is important to remember in using these equations to convert all weight units to grams and all volume units to liters.

$$n = \frac{d*x*V}{M} \tag{2-1a}$$

$$c = \frac{n}{V} \tag{2-1b}$$

where:
c = molar concentration of diluted acid or base (gm-moles/liter)
d = density of solution (gm/liter)
M = molecular weight of pure acid or base (gm/gm-mole)
n = number of gm-moles of pure acid or base
V = volume of solution (liters)
x = weight fraction of pure acid or base in solution

Molal units are used predominately by engineers and scientists studying electrolytes because the concentration calculation is independent of density and, hence, temperature. Molal concentration is the number of gm-moles per 1000 gm of solvent. Equations (2-2a) and (2-2b) show how to calculate and convert the molal concentration to molar concentration. Note that as the acid or base weight fraction approaches one (the grams of solvent approaches zero), the molal concentration approaches infinity.

$$m = \frac{1000*n}{d*(1-x)*V} \tag{2-2a}$$

$$c = \frac{d*(1-x)}{1000} * m \tag{2-2b}$$

c = molar concentration of diluted acid or base (gm-moles/liter)
d = density of solution (gm/liter)
m = molal concentration of diluted acid or base (gm-moles/kgm solvent)
n = number of gm-moles of pure acid or base
V = volume of solution (liters)
x = weight fraction of pure acid or base in solution

Normality units are predominately used in the charge-balance equations for acid and base ions. Normality concentration is the gram-ions of replaceable hydrogen or hydroxyl groups per liter of solution. A shorter notation of gram-equivalents per liter is frequently used. Equations (2-3a) and (2-3b) show how to calculate the normality concentration and how to convert from normality units to molar units. Table 2-1 shows the normality versus weight percent for some common reagents. Note that normality is not proportional to weight percent for a given reagent because the density of the solution also changes.

$$N = \frac{z^*d^*x}{M} \tag{2-3a}$$

$$c = \frac{N}{z} \tag{2-3b}$$

where:
d = density of the solution (gm/liter)
M = molecular weight of the acid or base (gm/gm-moles)
N = normality concentration of the diluted acid or base (gm-ions of replaceable hydrogen or hydroxyl groups/liter of solution)
x = weight fraction of pure acid or base in solution
z = number of replaceable hydrogen or hydroxyl groups per gm-mole

Reagent	**Wt %**	**Normality**
Hydrochloric acid	32	10.17
(z = 1)	38	12.35
Sulfuric acid	62.2	19.5
(z = 2)	77.7	27.2
	93.2	35.2
	98.0	36.0
Sodium hydroxide	10	2.75
(z = 1)	~~20~~ 25%	7.93
	50	19.1
Calcium hydroxide	5	1.36
(z = 2)	10	2.78
	15	4.30

Table 2-1. The normality of a reagent is not proportional to the weight percent.

As previously mentioned, it is desirable to use a ratio of reagent to influent flow for the abscissa of the titration curve for pH control system analysis. The method followed in this text will be to convert the concentration of the acid or base from the reagent, or influent streams in the effluent stream in molar, molal, and normality units to weight fractions via Eqs. (2-1a), (2-2a), or (2-3a) (solve the equations for x). It is necessary to eliminate composition dynamics in the use of the titration curve since it is a steady-state plot. Toward this purpose, it helps to visualize the influent and reagent streams combining in a pipeline instead of a tank so that there is an immediate translation from a change in reagent or influent flow to a change in the weight fractions of the effluent stream. For pure (undiluted) acid or base reagent and influent streams, the ratio of the reagent to influent flow is equal to the ratio of the weight fractions of the acid and base in the effluent stream. The weight fraction of the acid or base in the individual influent and reagent streams is used to calculate the diluted stream flows needed for control valve sizing and feed-forward calculations via Eqs. (2-4a) and (2-4b). In using weight fractions, it is necessary to designate which stream is the source of the acid or base and which stream is the solution. To reduce confusion in Eqs. (2-1) through (2-4), the subscript "rr" will designate a reagent stream source and a reagent stream solution (dilution); "re" will designate a reagent stream source and an effluent stream solution; "ii" will designate an influent stream source and an influent stream solution (dilution); and "ie" will designate an influent stream source and an effluent stream solution throughout the text. Since the influent flow is usually known before the effluent flow, Eq. (2-4b) is substituted for Eq. (2-4a) to yield Eq. (2-4c) for calculating the reagent flow. The desired flow ratio for the abscissa of the titration curve is the ratio of the diluted reagent to the diluted influent mass flow, which—per Eq. (2-4c)—is equal to the ratio of the effluent weight fractions multiplied by the inverse of the ratio of the dilution weight fractions. To convert from mass flow (pph) to volumetric flow (gpm), each mass flow must be divided by the diluted stream density in pounds per gallon. Note that most reagent concentrations are given in weight percent that must be divided by 100 to get the dilution weight fraction. If the sample was titrated in the laboratory, the initial sample volume is specified, the volume of reagent titrated to reach the pH setpoint is specified, and the reagent concentration used in the laboratory and plant are equal, then Eq. (2-4d) can be used to calculate the volumetric flow rate. The reagent volumetric flow can be converted to mass flow by multiplication by the reagent density. Equation (2-4e) shows simply that effluent stream mass flow is equal to the sum of the diluted influent and reagent stream mass flows.

$$F_r = \frac{X_{re}}{X_{rr}} * F_e \qquad (2\text{-}4a)$$

$$F_i = \frac{X_{ie}}{X_{ii}} * F_e \qquad (2\text{-}4b)$$

$$F_r = \frac{x_{re}}{x_{ie}} * \frac{x_{ii}}{x_{rr}} * F_i \quad (2\text{-}4c)$$

$$Q_r = \frac{V_r}{V_i} * Q_i \quad (2\text{-}4d)$$

$$F_e = F_r + F_i \quad (2\text{-}4e)$$

where:
F_e = effluent flow (pph)

F_i = influent mass flow (pph)

F_r = reagent mass flow (pph)

Q_i = influent volumetric flow (pph)

Q_r = reagent volumetric flow (pph)

V_i = influent sample volume (milliliters)

V_r = reagent volume titrated to reach pH setpoint (milliliters)

x_{ie} = weight fraction of influent acid or base in effluent stream

x_{ii} = weight fraction of influent acid or base in influent stream

x_{re} = weight fraction of reagent acid or base in effluent stream

x_{rr} = weight fraction of reagent acid or base in reagent stream

2-2. Ion Activity

As mentioned in the introduction, pH is more indicative of hydrogen ion activity than hydrogen ion concentration. Ion activity is the ratio of the escaping tendency of the component in solution to that at a standard state. Equation (2-5) shows that the ion concentration multiplied by an activity coefficient is equal to the ion activity. The activity units depend on the concentration units. The activity coefficient usually decreases from unity as the ion concentration increases from zero. In dilute solutions, the ions are far enough apart that the interaction between ions is negligible. Ion activity decreases as ion interaction increases. The change in activity is greater for ions with a large number of charges. As the concentration increases, the activity for some ions goes through a minimum and then increases due to ions grouping together. The activity of an ion also depends on the dielectric constant and temperature of the solvent and the concentration of the other ions in the solution. Table 2-2 lists the activity coefficients for some common ions in water for different ionic strengths (for these relatively dilute solutions, the ionic strength is equal to the sum of one half of the product of the individual ion concentrations and their ion valence or charge squared). Most effluent streams are dilute enough that the activity coefficient for hydrogen can be considered equal to one. However this is not true for many reagent and some influent streams.

$$a = y^{*}c \tag{2-5}$$

$$pH = -\log a_H \tag{2-6}$$

where:
a = ion activity
c = ion concentration
y = activity coefficient

Ion Type	Ion Size (Angstroms)	Ion Valence (charge)	Ionic Strength in Water at 25°C			
			0.005	0.010	0.050	0.100
Hydrogen	9	1	0.933	0.914	0.860	0.830
Lithium	6	1	0.929	0.907	0.835	0.800
Sodium	4.5	1	0.928	0.902	0.820	0.775
Hydroxyl	3.5	1	0.926	0.900	0.810	0.760
Potassium	3	1	0.925	0.899	0.805	0.755
Ammonium	2.5	1	0.924	0.898	0.800	0.750
Magnesium	8	2	0.755	0.690	0.520	0.450
Calcium	6	2	0.749	0.675	0.485	0.405
Carbonate	4.5	2	0.741	0.663	0.450	0.360
Sulfate	4	2	0.740	0.660	0.445	0.355
Phosphate	4	3	0.510	0.405	0.180	0.115

Table 2-2. The activity coefficient depends on the solvent, temperature, ion size, ion valence, and the ionic strength of the solution.

Equation (2-6) shows the definition of pH in terms of hydrogen ion activity instead of concentration. It is unwise to attach too much importance to pH measurement as an absolute indicator of hydrogen ion activity because the measurement is referenced to a measurement with a hydrogen electrode in a water solution (see Unit 4) that is not corrected for temperature and the activity of a single ion is difficult to obtain theoretically. The pH measurement provides a repeatable experimental rather than a true indication of hydrogen ion activity in water solutions. The pH measurement is in general not an indicator of hydrogen ion activity for pure acids or for solvents other than water. For a more complete discussion of the limitations of the pH measurement scale, the reader is directed to Ref. 1.

2-3. Ion Dissociation

An acid is a molecule that yields a hydrogen ion when it dissociates (breaks apart into its component ions) as shown in Eq. (2-7a) and a base is a molecule that yields a hydroxyl ion when it dissociates as shown in Eq. (2-7b). Water acts as both an acid and a base because it yields both a hydrogen ion and hydroxyl ion upon dissociation as shown in Eq. (2-7c). Neutralization is the association of the hydrogen and hydroxyl ions to form water; it is designated by the reverse arrow in Eq. (2-7c). Forward and reverse arrows are shown in these equations to show that ion association as well as dissociation occurs to maintain an equilibrium between the concentration of the species on both sides of the equation. These acid and base definitions were developed based on aqueous (water) solutions. A more general definition of an acid as a proton donor and a base as a proton acceptor is needed for nonaqueous solutions. Since the visualization of these definitions is more difficult and data on nonaqueous pH measurement is scarce, these definitions will not be explored further. The practical problems in making and interpreting nonaqueous pH measurement will be discussed in the section on electrode error in Unit 4. For more information on the proton donor and acceptor definitions, consult Ref. 2.

$$HA \rightleftharpoons H^+ + A^- \qquad (2\text{-}7a)$$

$$BOH \rightleftharpoons B^+ + OH^- \qquad (2\text{-}7b)$$

$$H_20 \rightleftharpoons= H^+ + OH^- \qquad (2\text{-}7c)$$

where:
HA = acid molecule
BOH = base molecule
H_20 = water molecule

A^- = negative ion from dissociation of acid

B^+ = positive ion from dissociation of base

H^+ = hydrogen ion (proton)

OH^- = hydroxyl ion

Equations (2-7) through (2-10) show the dissociation of some of the more common acid and base reagents. Note that sulfuric acid and calcium hydroxide dissociate twice and yield ions with a double charge.

$HCL \rightleftharpoons H^+ + Cl^-$ (dissociation of hydrochloric acid) (2-7d)

$H_2SO_4 \rightleftharpoons HSO_4^- + H^+$ (first dissociation of sulfuric acid) (2-8a)

$HSO_4^- \rightleftharpoons SO_4^{2-} + H^+$
(second dissociation of sulfuric acid) (2-8b)

$CaOH_2 \rightleftharpoons CaOH^+ + OH^-$
(first dissociation of calcium hydroxide) (2-9a)

$CaOH^+ \rightleftharpoons Ca^{2+} + OH^-$
(second dissociation of calcium hydroxide) (2-9b)

$NaOH \rightleftharpoons Na^+ + OH^-$ (dissociation of sodium hydroxide) (2-10)

A dissociation constant is used to define the relationship between the activities of the components in equilibrium with each other for each dissociation. It also provides a measure of the strength of the acid or base. In Sec. 2-4, the stream concentration will be assumed to be dilute enough so that the charge balance equations that utilize these dissociation constants will be simplified by the omission of activity coefficients. This assumption is usually valid for effluent streams, which are the streams of greatest interest in pH control loop performance analysis. The dissociation constant "K" typically falls numerically in the same range as the hydrogen ion concentration so that it is convenient to express it as a negative base ten logarithm like pH where the small p designates the power function, Eq. (2-11).

$$pK = -\log K \qquad (2\text{-}11)$$

Equations (2-12) through (2-15) show the relationship between the species concentrations and dissociation constant for the dissociation of the common reagents shown in Eqs. (2-7) through (2-10). Square brackets around each species denote that the quantity is an activity, which for dilute solutions is equal to the concentration in normality units. The numerator is the product of the ion concentrations while the denominator is the unionized acid or base concentration. The double subscript is used to designate the source and order of dissociation. A small "a" subscript designates the source is an acid and a small "b" subscript designates the source is a base. Frequently in references the small "a" and "b" subscripts are omitted. All such dissociation constants are for the hydrogen ion even though the source may be a base. The ionic product constant for water defined in Eq. (2-16) can be used to convert from a base to an acid dissociation constant by solving this equation for the hydroxyl ion and substituting it for the hydroxyl ion in the relationship for the base dissociation constant. The net result is the relationship between the dissociation constants shown in Eq. 2-17. Table 2-3a lists the pKa for some acids and Table 2-3b lists the pKb for some bases at 25°C (pKw = 14). The molecular weight is also shown because it is used in molar, molal, and normality concentration calculations. For more information on dissociation constants for aqueous and nonaqueous solutions, the reader is directed to Ref. 3.

It is important to realize that dissociation constants vary with temperature and hence the pH of the solution varies with temperature. A common misconception is that the temperature compensator in a pH measurement circuit corrects for this variation. Such compensators correct for the change in millivolt per pH unit relationship per the Nernst equation. Leeds & Northrup is the only manufacturer to date that has added to its pH controller the ability to specify a process pH versus temperature relationship. The more powerful distributed system controllers have the exponentials and polynomials as programmable function steps to calculate the correction to the setpoint or measurement.

The acid or base strength increases as its pK decreases. A pKa or pKb less than zero means that the product of the ion activities is greater than the activity of the source molecule. Such acids and bases are called "strong" since they are completely dissociated in the 0 to 14 pH range. Correspondingly, acids with a pKa or pKb greater than one are called "weak"—the relative weakness increases as the pKa or pKb increases. Note that the first dissociation constant of sulfuric acid classifies it as a "strong" acid while the second classifies it as a "weak" acid. In fact, sulfuric acid behaves like a mixture of a "strong" and "weak" acid. When

Acid Type	pK_{a1}	pK_{a2}	Molecular Weight
Acetic acid	4.75	—	60.05
Acrylic acid	4.26	—	72.06
Carbon dioxide	6.35	10.3	44.01
Fumaric acid	3.02	4.38	116.1
Hydrogen chloride	−6.1	—	36.46
Hydrogen cyanide	9.20	—	27.06
Hydrogen fluoride	3.17	0.60	20.01
Hydrogen sulfide	6.97	12.9	34.08
Maleic acid	1.94	6.22	116.1
Nitric acid	1.44	—	117.1
Phosphorous acid	2.00	6.40	82.00
Phthalic acid	2.80	5.10	166.1
Sorbic acid	4.77	—	112.1
Sulfuric acid	−3.0	1.99	98.08

Table 2-3a. These acid dissociation constants are for aqueous solutions at 25°C.

Base Type	pK_{b1}	pK_{b2}	Molecular Weight
Ammonia	4.76	—	17.03
Calcium hydroxide	1.40	2.43	74.09
Sodium hydroxide	−0.8	—	40.01
Urea	13.9	—	60.06

Table 2-3b. These base dissociation constants are for aqueous solutions at 25°C.

Temperature (degrees C)	pK_w
0	14.94
5	14.73
10	14.54
15	14.35
20	14.17
25	14.00
30	13.83
35	13.68
40	13.54
45	13.40
50	13.26

Table 2-4. The ionic activity product of water depends on temperature.

the pH equals the pKa, the negative acid ion activity is equal to the acid molecule activity. The dissociation is at the midpoint. This midpoint can be spotted on the titration curve if the dissociation constants are not too close together and can be used as a flag for identifying the dissociation constants from laboratory titration curves for signal characterization and control system simulation programs. This technique will be discussed in greater detail in Units 3 and 7.

$$K_a = \frac{[H^+][Cl^-]}{[HCl]}$$ (dissociation constant for hydrochloric acid) (2-12)

$$K_{a1} = \frac{[H^+][HSO_4^-]}{[H_2SO_4]}$$ (first dissociation constant for sulfuric acid) (2-13a)

$$K_{a2} = \frac{[H^+][SO_4^{2-}]}{[HSO_4^-]}$$ (second dissociation constant for sulfuric acid) (2-13b)

$$K_{b1} = \frac{[OH^-][CaOH^+]}{[Ca^2 OH]}$$ (first dissociation constant for calcium hydroxide) (2-14a)

$$K_{b2} = \frac{[OH^-][Ca^{2+}]}{[CaOH^-]}$$ (second dissociation constant for calcium hydroxide) (2-14b)

$$K_b = \frac{[OH^-][Na^+]}{[NaOH]}$$ (dissociation constant for sodium hydroxide) (2-15)

$$K_w = [H^+][OH^-]$$ (ionic product constant for water) (2-16)

$$pK_b = pK_w - pK_a$$ (2-17)

2-4. Ion Charge Balance

The rest of Unit 2 may be skipped if the reader is not interested in the equations used for programming signal characterization calculations or simulating pH processes and titration curves. These equations are not used for pH control loop analysis or implementing conventional feedback control. The understanding of these equations is also not necessary for implementing preprogrammed pH feedforward control or linearization or generating titration curves by use of the program in Appendix E.

Since solutions are neutral, the existence of just one type of ion in solution is not possible. Thus, pH measurement cannot be stated to be a measurement of just the hydrogen ion. In water solutions, hydrogen ions will be accompanied by hydroxyl ions per the ionic product shown in Eq. 2-16. If no other ions are in the solution, charge neutrality demands that the hydrogen ion concentration equal the hydroxyl ion concentration. If other ions are in solution, the sum of each ion concentration multiplied by its charge (normality) must equal zero. This is fortunate because it permits the use of an interval-halving search method to iteratively search for the pH for a given set of ion concentrations that makes the charge balance zero. The interval-halving search method is relatively fast and simple. The excess charge is calculated for the pH guess at the midpoint of the pH search interval. If the excess charge is negative, the lower pH search limit is increased to the midpoint pH. If the excess charge is positive, the upper pH search limit is decreased to the midpoint pH. This interval halving continues until the interval is less than a specified allowable error. If the pH and concentrations of all but one acid or base in a solution are known, the unknown acid or base concentration can be solved for directly from the charge balance equation.

The charge balance equation is set up by summing all the normalities of all the ions in solution and setting the sum equal to zero. The acid or base concentrations and not the ion concentrations are generally given. A strong acid or base is completely ionized so that the ion concentration is equal to the acid or base concentration. A weak acid or base is only partially ionized. The concentration of the ions can be calculated from the relationships for the dissociation constants and combined in an expression based on the acid or base concentration. Equations (2-18) through (2-20) show the expressions for single, double, and triple dissociations, respectively, of a weak acid or base. For the derivations of the expressions for single and double dissociations, the reader is directed to Ref. 4. A derivation of the expression for the third dissociation can be obtained by extending the same method. Note that the sign of the parameter "s" determines whether the expression is for a weak acid or base. Also, the base dissociation constants pKb must be converted to acid dissociation constants pKa via Eq. (2-17) to use the expressions for bases. If the pH is two units or more larger than the pKa for acids or two or more units less than the pKa for bases, Eqs. (2-21) through (2-23) contribute less than 0.01 when summed with 1 in Eqs. (2-18) through (2-20). The numerators and denominators are then approximately equal to 1 and Eqs. (2-18) through (2-20) reduce to simply $s*N$. Thus, the weak acids can be considered completely ionized for a pH greater than the pKa by 2 and weak bases can be considered completely ionized for a pH less than the pKa by 2. Weak acids behave like strong acids at a high pH and weak bases behave like strong bases at a low pH. The contribution of each weak acid or weak base to the charge balance equation is represented by one of Eqs. (2-18) through (2-20). Equation (2-24) shows the charge balance equation for a strong acid and base and Eq. (2-25) shows the charge balance equation for a weak acid and base. The last two terms in both equations are for the hydrogen and hydroxyl ion concentrations.

$$N_1 = \frac{1}{(1 + P_1)} * s * N \quad (2\text{-}18)$$

$$N_2 = \frac{(1 + 0.5 * P_2)}{(1 + P_2 * (1 + P_1))} * s * N \quad (2\text{-}19)$$

$$N_3 = \frac{(1 + 0.33 * P_3 * (2 + P_2))}{(1 + P_3 * (1 + P_2 * (1 + P_1)))} * s * N \quad (2\text{-}20)$$

$$P_1 = 10^{(s * (pH - pK_1))} \quad (2\text{-}21)$$

$$P_2 = 10^{(s * (pH - pK_2))} \quad (2\text{-}22)$$

$$P_3 = 10^{(s * (pH - pK_3))} \quad (2\text{-}23)$$

$$N_b - N_a + 10^{-pH} + 10^{(pH - pK_w)} = 0 \quad (2\text{-}24)$$

$$\frac{1}{(1 + P_{1b})} * N_b - \frac{1}{(1 + P_{1a})} * N_a + 10^{-pH} + 10^{(pH - pK_w)} = 0 \quad (2\text{-}25)$$

where:

N = concentration of acid or base (normality)

N_1 = concentration of ions from a single dissociation (normality)

N_2 = concentration of ions from a double dissociation (normality)

N_3 = concentration of ions from a triple dissociation (normality)

N_a = concentration of acid (normality)

N_b = concentration of base (normality)

pK_1 = first dissociation constant

pK_2 = second dissociation constant

pK_3 = third dissociation constant

pK_w = ionic product for water

s = ion sign (s= −1 for acids and s= +1 for bases)

Exercises

2-1. *If an influent stream has a specific gravity of 1.1 and a normality of 2.75 due to the presence of sodium hydroxide, what is the weight percent of sodium hydroxide in the influent?*

2-2. *If the reagent contains 20% by weight hydrochloric acid, what flow in pph of reagent is required to neutralize 100 pph of the influent described in Exercise 1-1 for an effluent specific gravity of 1.08.*

2-3. *At 7 pH does the concentration of acid equal the concentration of base in normality units for a strong acid and base system?*

2-4. *At 7 pH does the concentration of acid equal the concentration of base in normality units for a weak acid and base system?*

2-5. *If the temperature of a solution is 50°, at what pH does the hydrogen concentration equal the hydroxyl concentration in normality units?*

References

[1]Bates, R. G., *Determination of pH Theory and Practice*, John Wiley & Sons, 1964, pp. 17-33.
[2]Moore, W. J., *Physical Chemistry*, 3rd Edition, Prentice Hall, 1962, pp. 361-362.
[3]*Lange's Handbook of Chemistry*, 12th Edition, McGraw-Hill Book Company, 1979, pp. 5-13, 5-47.
[4]Shinskey, F. G., *pH and pION Control in Process and Waste Streams*, John Wiley & Sons, 1973, pp. 54-67.

Unit 3:
pH Titration Curves

UNIT 3

Titration Curves

This unit demonstrates how the titration curve changes with composition, presents the alternate methods of generating the titration curve, and gives an overview of the more important uses of the titration curve.

Learning Objectives—When you have completed this unit you should:

A. **Be able to classify the pH system as to general type based on the titration curve.**

B. **Know how to get titration curves that are useful for system design.**

C. **Recognize the potential uses of the titration curve.**

3-1. Characteristics

The control system sensitivity and the pH process gain increase as the slope of the titration curve increases. While the details on assessing the exact implications of this statement are covered in subsequent units, it is appropriate to now discuss how the slope changes value and location as a function of stream composition. Figures 3-1 through 3-4 show how the titration curve changes for the four possible combinations of strong and weak acids and bases. The strong acid pKa and the strong base pKb are both equal to 0 pH and the weak acid pKa and the weak base pKb are both equal to 3 pH.

The strong acid and strong base curve shown in Fig. 3-1 is distinguished by its vertical slope throughout most of the pH scale range. Figure 1-1b (in the introduction) showed that this slope is not vertical but changes by a factor of 10 for each pH unit deviation from 7 pH for a solution temperature of 25°C. Titration curves rarely consist of straight lines; the blowup of a straight line usually will reveal another curve. A titration curve without a clearly defined abscissa is worthless because the shape will change with the abscissa range. The neutral point occurs at a pH equal to one half of the pKw so that its location depends on solution temperature. The equivalence point coincides with the

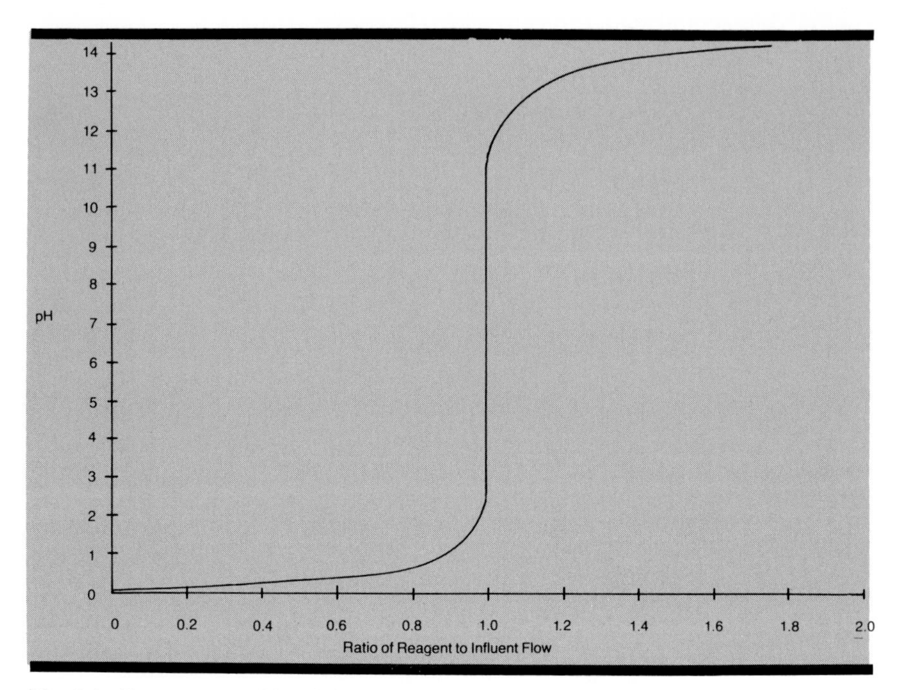

Fig. 3-1. For a strong acid and strong base system, the control system sensitivity and pH process gain is a concern for any pH setpoint.

neutral point for a strong acid and strong base system. The titration curve slope is steepest at the equivalence point. The definitions of the neutral and equivalence points are given to avoid confusion.

The *neutral point* on the titration curve is the point at which the hydrogen ion concentration equals the hydroxyl ion concentration.

The *equivalence point* on the titration curve is the point at which the acid ion concentration equals the base ion concentration.

While the strong acid and strong base system appears to be symmetrical about the equivalence point, the curve is only truly symmetrical if the dissociation constants of both are equal to zero for a temperature of 25°C. The midpoint of the lower flat portion of the titration curve occurs at the pH equal to the pKa for the strong acid. The midpoint of the upper flat portion(s) occur at the pH equal to the pKw for water and the pKa (not pKb) for the strong base. Two distinct upper flat portions occur if the pKw and pKa are sufficiently separated. However, the exact shape is difficult to obtain due to the large electrode error and activity coefficient error at these extremes in concentration and is unimportant since it lays outside the practical pH measurement and setpoint range.

The strong acid and strong base system has a maximum slope and range of steep slopes greater than any other pH system. The control system sensitivity and the pH process gain is a concern for any pH setpoint other than at the extremes of the pH scale range.

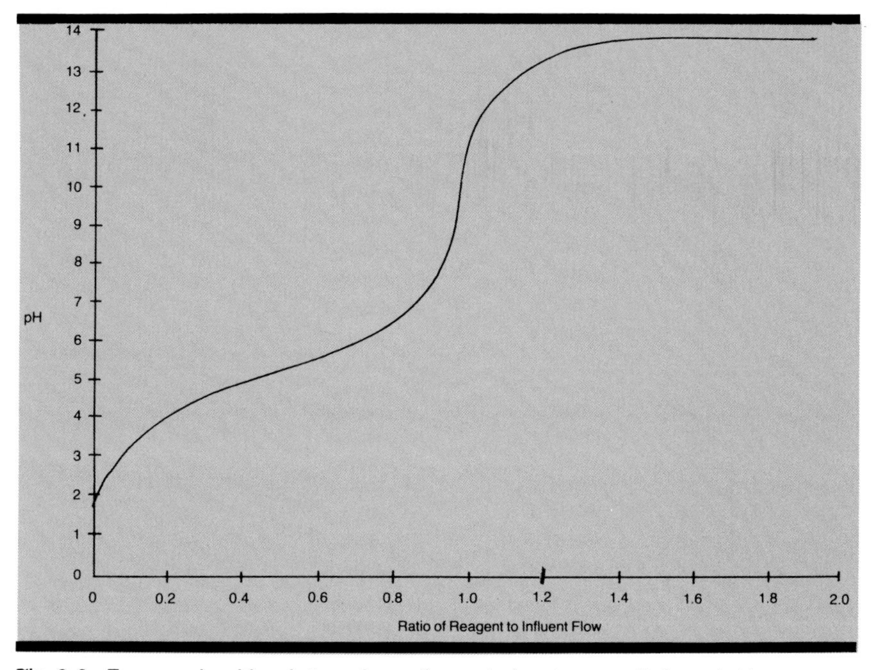

Fig. 3-2. For a weak acid and strong base, the control system sensitivity and pH process gain is a concern for high pH setpoints.

The weak acid and strong base system in Fig. 3-2 is distinguished by its steep slope in the upper pH scale range. The center of this steep slope is the equivalence point location dependent upon the dissociation constant of the weak acid. The control system sensitivity and pH process gain is a concern for pH setpoints greater than 1 pH unit above the weak pKa and 1 pH unit below the strong base pKa or the water Pkw. The sensitivity and gain is least for a pH setpoint equal to the pKa, which is at the midpoint of the lower flat portion. The flat portions at the upper and lower ends of the pH range are outside the practical pH setpoint range.

The strong acid and weak base system shown in Fig. 3-3 is distinguished by its steep slope in the lower pH scale range. The center of this steep slope is the equivalence point location dependent upon the dissociation constant of the weak base. The control system sensitivity and pH process gain is a concern for pH setpoints greater than 1 pH unit below the weak base pKa and 1 pH unit above the strong acid pKa. The sensitivity and gain is least for a pH setpoint equal to the pKa, which is at the midpoint of the upper flat portion. The flat portions at the upper and lower ends of the pH range are outside the practical pH setpoint range.

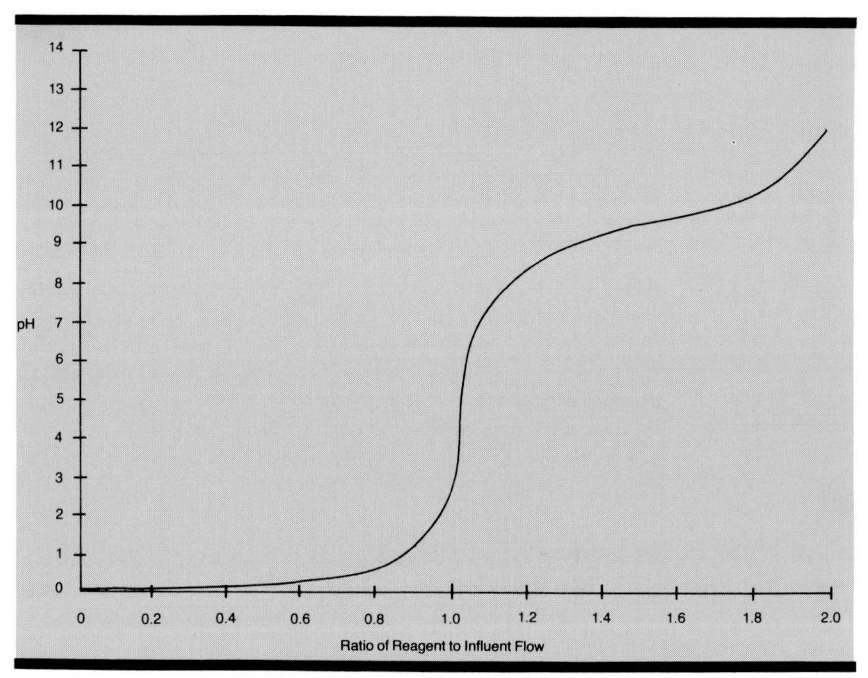

Fig. 3-3. For a strong acid and weak base, the control system sensitivity and pH process gain is a concern for low pH setpoints.

The weak acid and weak base system in Fig. 3-4 is distinguished by the lack of a steep slope throughout the range. The titration curve slope is greatest at the equivalence point but is still relatively small compared to the previous systems. The equivalence point ordinate depends upon both the acid and base dissociation constants. While the magnitude of the pH process gain is low, the gain is still nonlinear. The system should not be considered as easy to control.

A common misconception is that a titration curve consists of a single S-shaped curve. In Fig. 3-5, a weak acid with pKa dissociation constants at 2.5 And 5.5 pH is titrated with a weak base with pKb dissociation constants at 8.5 and 11.5 pH. Notice that three S-shaped titration curves are formed with an equivalence point at the point of steepest slope for each S. The dissociation

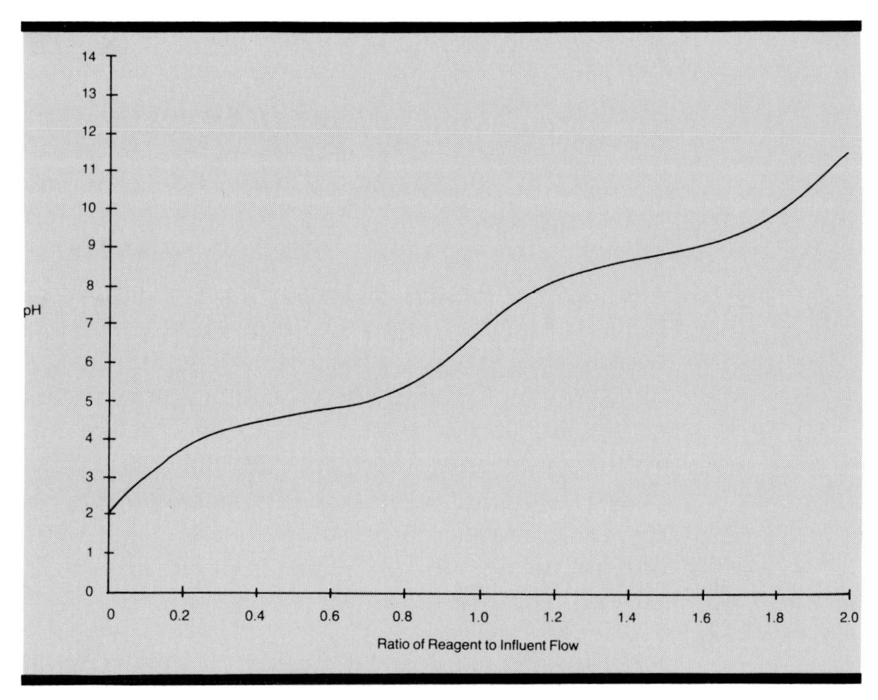

Fig. 3-4. For a weak acid and weak base, the control system sensitivity and pH process gain is of relatively less concern even at the equivalence point.

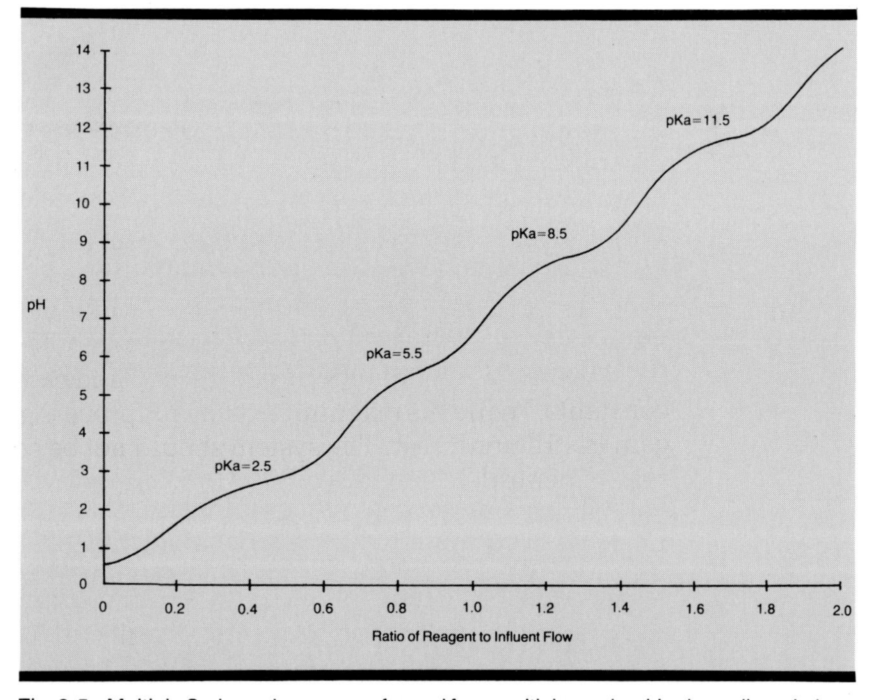

Fig. 3-5. Multiple S-shaped curves are formed from multiple weak acid or base dissociation constants if they are spread apart by more than two pH units.

constants have to be more than 2 pH units apart to create multiple S-shaped curves. The curves are symmetrical because the concentrations and the distances between the dissociation constants are equal. Normally pH titration curves are not symmetrical.

3-2. Laboratory Generation

Manually generated laboratory titration curves are typically done by placing a known volume of the process sample in a beaker, adding a fixed volume of reagent from the burette, and logging the pH indication. The data points are then manually plotted and a smooth curve is sketched between the points. The volume of reagent added between the data points must be reduced drastically near the equivalence point. If the sample or reagent contains a strong acid or base, it is difficult to generate data points near the equivalence point. The abscissa is usually labeled with the volume of reagent added.

If the titration curve has a long, flat tail at low or high pH, data points may not be plotted until the curve starts to bend. In order to estimate the valve rangeability and hysteresis requirement and the mixing equipment size and agitation requirements, the starting pH point, which represents the influent pH, must be accurately plotted.

If the sample or the reagent contains solids, the time required for the solids to dissolve and the ions to get into solution to establish the solution ion balance can take from several seconds to several minutes. If the pH has not reached its final value before the reagent volume is incremented, the titration curve become distorted. If a gaseous reagent such as ammonia is bubbled into the sample, there is dissolution time to account for; also, bubbles may rise to the surface and escape from the solution before dissolving so that the amount of reagent added is not equal to the amount in solution.

If the volume of the sample is given and the concentration of the reagent used in the laboratory and the control system are equal, the reagent volumetric flow can be calculated for a given influent flow and pH setpoint per Eq. (2-4d). The concentration of the laboratory and field reagents must be known to use the titration curve for pH control system design.

The laboratory is usually heated, vented, and air conditioned so that the sample temperature during titration is rarely equal to the ambient temperature or process temperature. The sample pH will change with temperature since the dissociation constants and water ionic product change with temperature. The sample temperature should be maintained equal to the process temperature at the electrodes. If the process is cooled or heated to bring its temperature within the permissible range for the electrodes, a temperature sensor should be installed at the electrodes so that the sample can be titrated at the same temperature. If the process temperature at the electrodes changes by more than a few degrees, multiple samples should be titrated at different temperatures to determine the change in pH with temperature near the pH setpoint.

Samples should be stored in Pyrex glass or polyethelyne plastic bottles. Soft glass bottles should not be used because the glass solubility is great enough to raise the pH due to the absorption of alkali ions by the sample. Cork stoppers should not be used because the corks are highly acidic. Samples exposed to the air can absorb enough carbon dioxide to lower the pH. This is a particular problem for nearly pure water and caustic samples. The pH of absolutely pure water at 25°C can change from a pH of 7 to a pH of 5.7 by exposure to air. The slope of caustic solutions will decrease in the region of 6 to 7 pH from the moderating effect of carbonates due to the absorption of carbon dioxide from the air. This moderating effect of carbonates helps to regulate the pH of many aquatic streams and reduces the difficulty of waste and raw water pH control. For more information on the effect of carbon dioxide on the pH of samples, the reader is directed to Refs. 1 and 2.

The errors at the high and low ends of the pH scale depend upon the type of pH measurement electrode used. This is particularly a problem for pH measurements above 10 pH when the sample or reagent contains the sodium ion or other large alkali ions. Unit 4 will quantify these errors.

If the titration curve changes with time, separate samples should be gathered over a representative period and individually titrated. The samples should not be combined for titration. The pH controller sees individual titration curves, not a composite. This as well as other common problems that reduce the utility of laboratory titration curves are listed below.

Common Problems with Laboratory Titration Curves

1. An insufficient number of data points was generated near the equivalence point.
2. The starting pH (influent pH) data point was not plotted.
3. The sample or reagent solids dissolution time effect on the abscissa was omitted.
4. The gaseous reagent dissolution time and escape effect on the abscissa was omitted.
5. The sample volume was not specified.
6. Reagent concentration was not specified.
7. The sample temperature during titration is different than the process temperature.
8. The influent sample was contaminated by absorption of carbon dioxide from the air.
9. The influent sample was contaminated by absorption of ions from the glass beaker.
10. The laboratory measurement electrode type was different than the field electrode type.
11. A composite sample instead of individual samples was titrated.
12. The reagent used for the titration curve was not the same type as that used by the control system.

Automatically generated titration curves are typically generated by a motor-driven syringe type of burette that is synchronized with a recorder chart drive. The titrator may be designed to emulate the manual titration where smaller reagent volumes are added near the equivalence point. This is accomplished by turning on the motor of the burette and the drive of the recorder and turning off the potentiometer and pen drive when the error between the old and the new pH is small and vice-versa when the error is large. Figure 3-6 shows this type of titrator.

Automatic titrators that are electronically equipped to compute the first and second derivative of pH with respect to time, and hence reagent flow, can automatically locate the equivalence point. The first derivative, which is the slope of the titration curve, reaches a maximum at the equivalence points. The second derivative changes sign at the equivalence points. A plot of first derivative helps in the control system design because it is indicative of the pH process sensitivity. Figure 3-7 shows a plot of the first derivatives for the weak acid and weak base titration curve in Fig. 3-5. For more information on automatic titrators, consult Ref. 3.

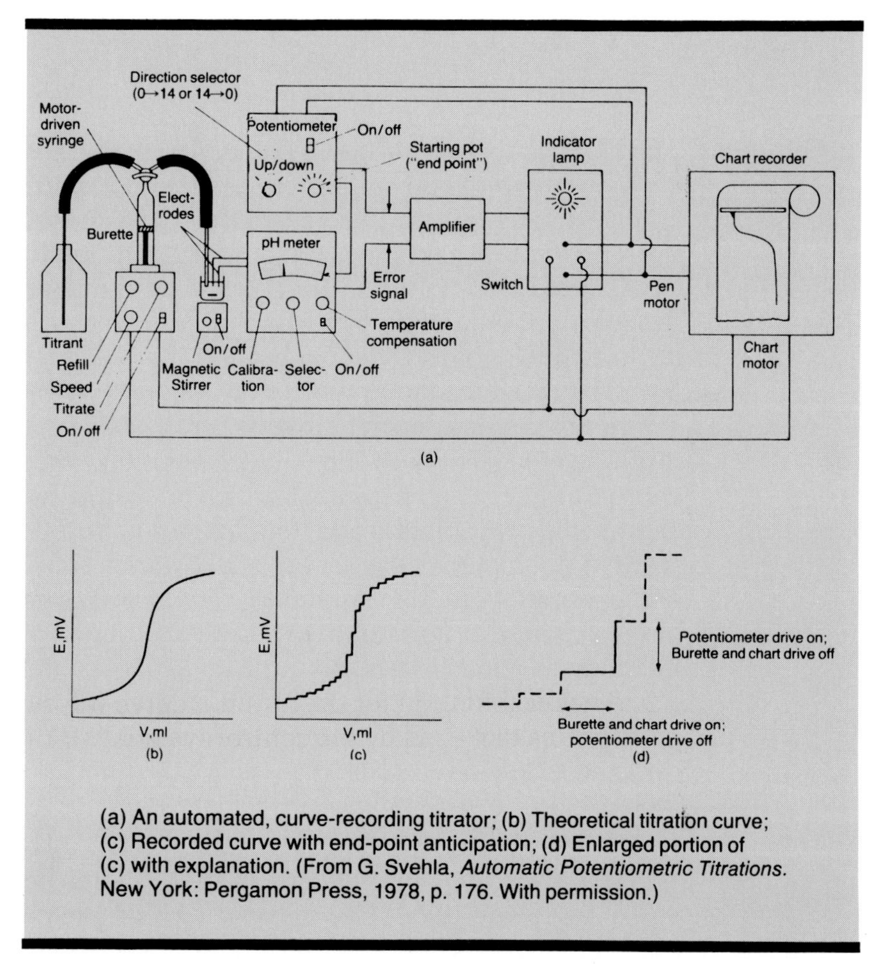

(a) An automated, curve-recording titrator; (b) Theoretical titration curve; (c) Recorded curve with end-point anticipation; (d) Enlarged portion of (c) with explanation. (From G. Svehla, *Automatic Potentiometric Titrations.* New York: Pergamon Press, 1978, p. 176. With permission.)

Fig. 3-6. The automatic titrator with reagent increment adjustment creates a staircase titration curve (Ref. 3).

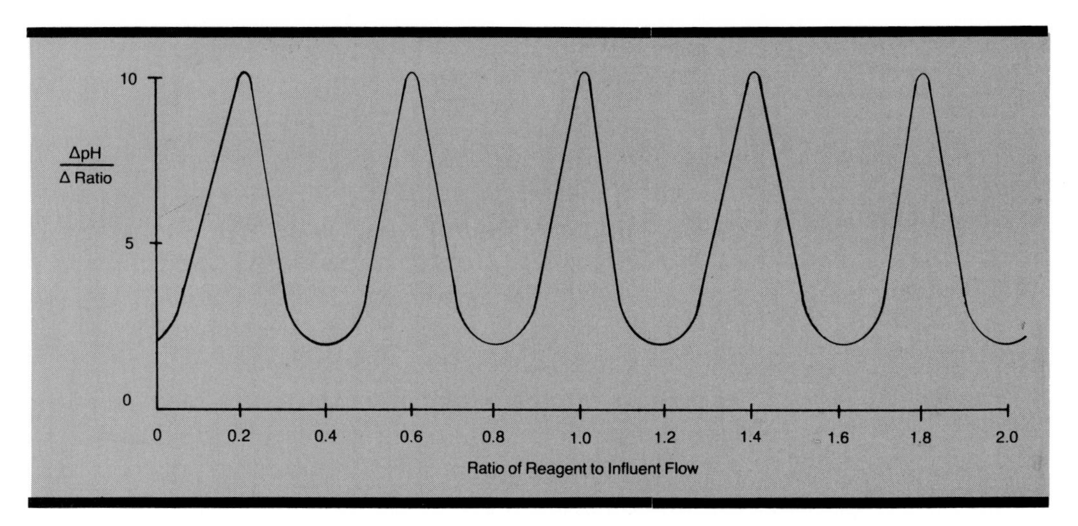

Fig. 3-7. The first derivative of the titration curve (Fig. 3-5) is indicative of the pH process sensitivity which reaches a maximum at the equivalence points.

3-3. Computer Generation

The FORTRAN subroutine in Appendix E solves the charge-balance equation and computes the pH and ion concentrations for any number of acids and bases that have up to three dissociation constants each. It does not have activity coefficients. It is designed to be called from another program that provides the proper input data. If the types and quantities of acids and bases are not known, or the effect of the activity coefficients is important, a Nelder-Mead program can be used to adjust the input data such as acid or base weight fraction or dissociation constant to minimize the error between it and a laboratory or more accurate computer-generated titration curve. The main advantage of the subroutine in Appendix E is its relative computational simplicity which allows it to be used in dynamic simulation programs without lengthy run times. The interval-halving search method is simple and relatively efficient—many other search methods develop instabilities due to the large concentration range covered for pH calculations. For more information on the problems with iterative search techniques for pH, the reader is directed to Ref. 4.

The Advanced Continuous Simulation Language (ACSL) available from Mitchell and Gauthier, Assoc., Inc., or the Continuous Simulation Modeling Program (CSMP) from IBM can be used to generate titration curves by use of the RAMP macro to ramp the reagent weight fractions while the influent weight fractions are held constant. The ACSL and CSMP plot functions facilitate plotting of the titration curves as printer plots or line plots. The effects of different influent and reagent compositions, which would be expensive to generate in the field, can be readily studied and the results used for control system design.

Dynamic simulation programs must have double precision instead of single precision arithmetic to show pH control of a strong acid and strong base within one pH of the equivalence point due to round-off error in the concentration dynamics. The concentration dynamics are simulated by integrators that maintain the mass balance of each species in the mixing equipment. Variable step integration methods develop problems due to noise from interval halving and the high sensitivity of the pH process. Fixed-step integration methods provide accuracy well within the limits of pH control in the field.

If a more accurate computer-generated titration curve is needed, the Equilibrium Composition of Electrolyte Solutions (ECES) program from OLI Systems, Inc., can be used for steady-state modeling of aqueous electrolyte systems. The program first generates a set of algebraic equations describing the equilibrium conditions from the user-defined chemistry and physical properties. The program then solves the equations simultaneously for the unknown values based upon a set of known values supplied by the user. The program may develop convergence problems for particularly difficult systems. While the program has a library of physical properties, such as activity coefficients, the user frequently has to find or estimate input data and regress the data with the program. Since the output data is in tabular form, separate graphics software such as TELAGRAF from Integrated Software Systems Corporation is used to generate line plots of the ECES program results. For more information on the capabilities and limitations of the ECES program, the reader is directed to Ref. 5.

3-4. Field Generation

The titration curve can be field generated by fixing the influent flow, slowly ramping a linear reagent valve, measuring the influent flow, and using a XY recorder to plot the pH measurement versus the reagent flow. A conventional trend recorder can be used if the speed is slow enough, the installed characteristic of the reagent valve linear enough, and the ramping stroke smooth enough for the desired accuracy of the titration curve. The pH

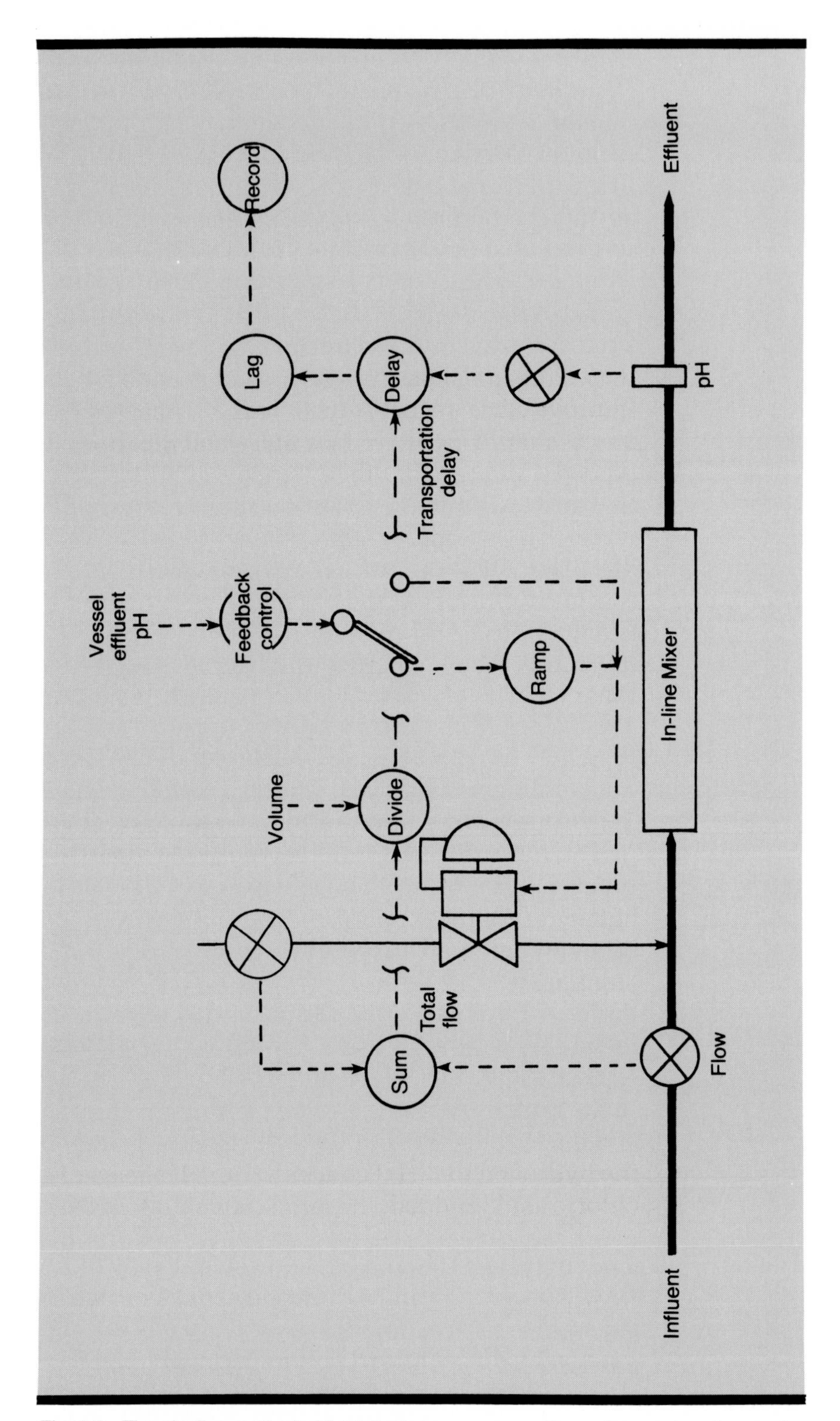

Fig. 3-8. The pipeline method of field titration curve generation reduces reagent usage for off-line titration and facilitates on-line titration.

measurement can be made in a mixed tank if the effluent flow is stopped so that the tank temporarily acts like a big beaker. The tank level should be low to minimize reagent usage for the titration. A better method used to reduce reagent consumption is to mix the reagent and influent in a pipeline prior to the tank and add a pH measurement to the pipeline. The pipeline method can be used to gather data points for updating the titration curve when the control system is on automatic. This on-line identification method only gathers points across a large portion of the titration curve during startup or during poor control. However, these additional data points are not needed for good control. To improve curve readability, the pH and reagent flow measurements should be filtered for noise and the flow measurement passed through a time delay set equal to the turnover time for the tank method or to the transport time for the pipeline method. Field curve identification is particularly useful for systems in which the signal linearization input data or controller mode settings need to be adapted to a variable titration curve. Field titration curves do not have the problems listed in Sec. 3-2 for laboratory titration curves, except for solids and gas dissolution time.

3-5. Buffering

Some titration curves show a long portion of relatively flat slope. The addition of reagent in this portion of the titration curve has little effect on the pH. This flatness of the curve can be due to buffering, which occurs for mixtures of a weak acid and its strong base salt or a weak base and its strong acid salt. Some of the common encountered buffer systems are acetic acid-acetate, carbonic acid-bicarbonates, and citric acid-citrates. Equation (3-1) is the expression of the dissociation constant for the acetic acid-acetate system solved for the hydrogen ion concentration. In the buffered region, the concentrations of acetate ions $[Ac^-]$ and the unionized acetic acid [HAc] are very large with respect to the hydrogen ion $[H^+]$ concentration. If a reagent such as hydrochloric acid is added, the large concentration of acetate ions quickly converts the extra hydrogen ions to unionized acetic acid. If a base such as sodium hydroxide is added, the extra hydroxyl ions neutralize hydrogen ions but these hydrogen ions are quickly replaced by the ionization of acetic acid. The percent changes in the unionized acetic acid and acetate ions is small because the magnitude of these concentrations is large. The ratio of acetic acid to acetate concentration shown in Eq. (3-1) is thus nearly constant so that the hydrogen ion concentration is also constant for a constant temperature and thus constant pKa. Buffered solutions are used for checking the accuracy of pH electrodes because the pH of the test solution remains constant even after moderate contamination.

$$[H^+] = \frac{[HAc]}{[Ac^-]} * pK_a \tag{3-1}$$

where:
Ac^- = acetate ion concentration (normality)
HAc = unionized acetic acid concentration (normality)
H^+ = hydrogen ion concentration (normality)
pK_a = dissociation constant for acetic acid

3-6. Uses

The titration curve is the single most important piece of information for the design, commissioning, and trouble shooting of pH control systems. The following uses will be discussed in subsequent units:

Titration Curve Uses

1. pH measurement range and accuracy specification
2. Reagent flow measurement rangeability and accuracy specification
3. Reagent control valve rangeability and hysteresis specification
4. Mixing equipment size, type, and agitation specification
5. Signal characterization input data specification
6. Controller proportional band, notch gain, and breakpoint selection

Exercises

3-1. *Rank the four major types of titration curves shown in Figs. 3-1 through 3-4 from least to most difficult for control for a 6 pH setpoint.*

3-2. What location of pH setpoint would be best for control for a weak acid with a single dissociation constant titrated with a strong base?

3-3. What problems could occur with a laboratory titration that has sodium hydroxide in the sample?

3-4. Will the interval-halving search method ever fail to converge on the pH that corresponds to a zero excess charge?

3-5. Why are laboratory titration curves more accurate than computer-generated curves for mixtures of many different ions at high concentrations?

3-6. Why can't the pipeline method of titration curve generation be used for lime reagent?

References

[1]*Modern pH and Chlorine Control,* Taylor Chemicals, Inc., 1945, pp. 83, 86.
[2]Shinskey, F. G., *pH and pION Control in Process and Waste Streams,* John Wiley & Sons, 1973, pp. 69-71.
[3]Skoog, D. A., and West, D. M., *Principles of Instrumental Analysis,* Saunders College, 1980, pp. 573-576.
[4]Richter, J. D. et. al., "Waste Neutralization Control—Digital Simulation Spots Nonlinearities," *Instrumentation Technology,* Vol. 21, No. 4, April 1974, pp. 35-40.
[5]Sanders, S. J., "Case Studies in Modeling Aqueous Electrolyte Solutions with the ECES and FRACHEM Programs" (to be published).

Unit 4: Measurement

UNIT 4

Measurement

This unit describes the principle of electrode operation, the source of errors, the causes of poor response, and the methods of installation in enough detail so that the reader does not have unreasonable expectations as to measurement performance and has a basis for designing to avoid and troubleshooting to correct pH measurement application problems.

Learning Objectives—When you have completed this unit you should:

A. Understand how the pH signal is generated.

B. Be able to correlate the source and symptoms of a measurement error.

C. Appreciate the importance of electrode assembly selection and installation on measurement and control error.

4-1. Electrode Description

Galvanic cells develop an electric potential as the result of a chemical reaction at the electrodes with the ions in an electrolyte solution. Two electrodes in contact with an electrolyte solution, each a half cell, connected to an external load, pH meter, or potentiometer are needed to provide a closed circuit and cause an electrical current to flow. A familiar example of a galvanic cell is the car battery that consists of lead and lead dioxide electrodes in contact with a dilute sulfuric acid solution.

The pH measurement and reference electrodes each have an internal galvanic half cell. The measurement and reference electrode half cells typically both consist of a silver wire with a silver chloride coating at its end immersed in a solution with chloride ions. The reference electrode solution is usually potassium chloride while the measurement electrode solution is usually a chloride buffer with a hydrogen activity approximately equivalent to 7 pH. Sometimes a calomel reference electrode is used, which consists of a platinum wire coated with mercury in contact with layers from top to bottom of a solution of potassium chloride saturated with mercurous chloride, a paste of potassium

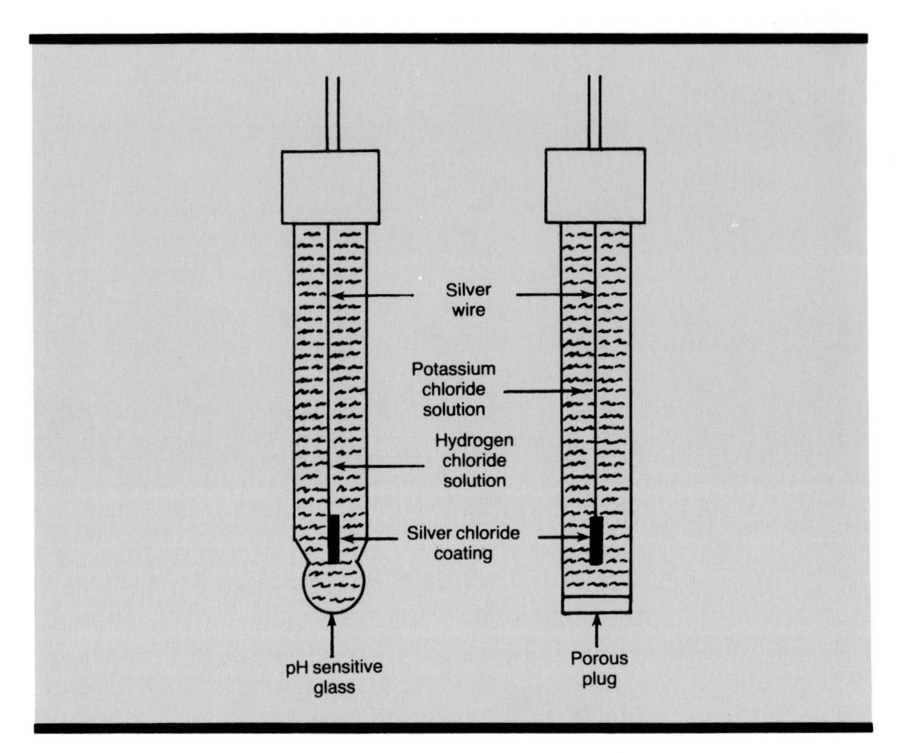

Fig. 4-1. The pH measurement and reference electrodes each have an internal galvanic half cell that will cancel out in the circuit if equal.

chloride, a paste of mercurous chloride, and finally a paste of mercury. The calomel half cell develops a potential that depends upon the concentration of the potassium chloride in the solution and changes more with temperature when the solution is saturated with potassium chloride. (Ref. 1). The use of a calomel (mercury-mercurous chloride) half cell in the reference electrode and a silver-silver chloride half cell in the measurement electrode requires a standardization potential to be set in the pH transmitter to cancel the difference between the two half cell potentials. In the early 1970s, "solid state" reference electrodes became popular because the internal gel was not as easily contaminated by external process fluid and the large, porous liquid junction was less likely to be completely coated. The internal gel was saturated with potassium chloride and silver chloride and a piece of wood, ceramic, or polymer were used as the porous medium at the tip of the electrode. (Ref. 2).

The Glasteel combination measurement and reference electrode developed and patented (U.S. patent 3,787,307) by Pfaudler is constructed from a fusion of glass and steel. The result is a higher temperature rating (140°C) and greater structural strength for

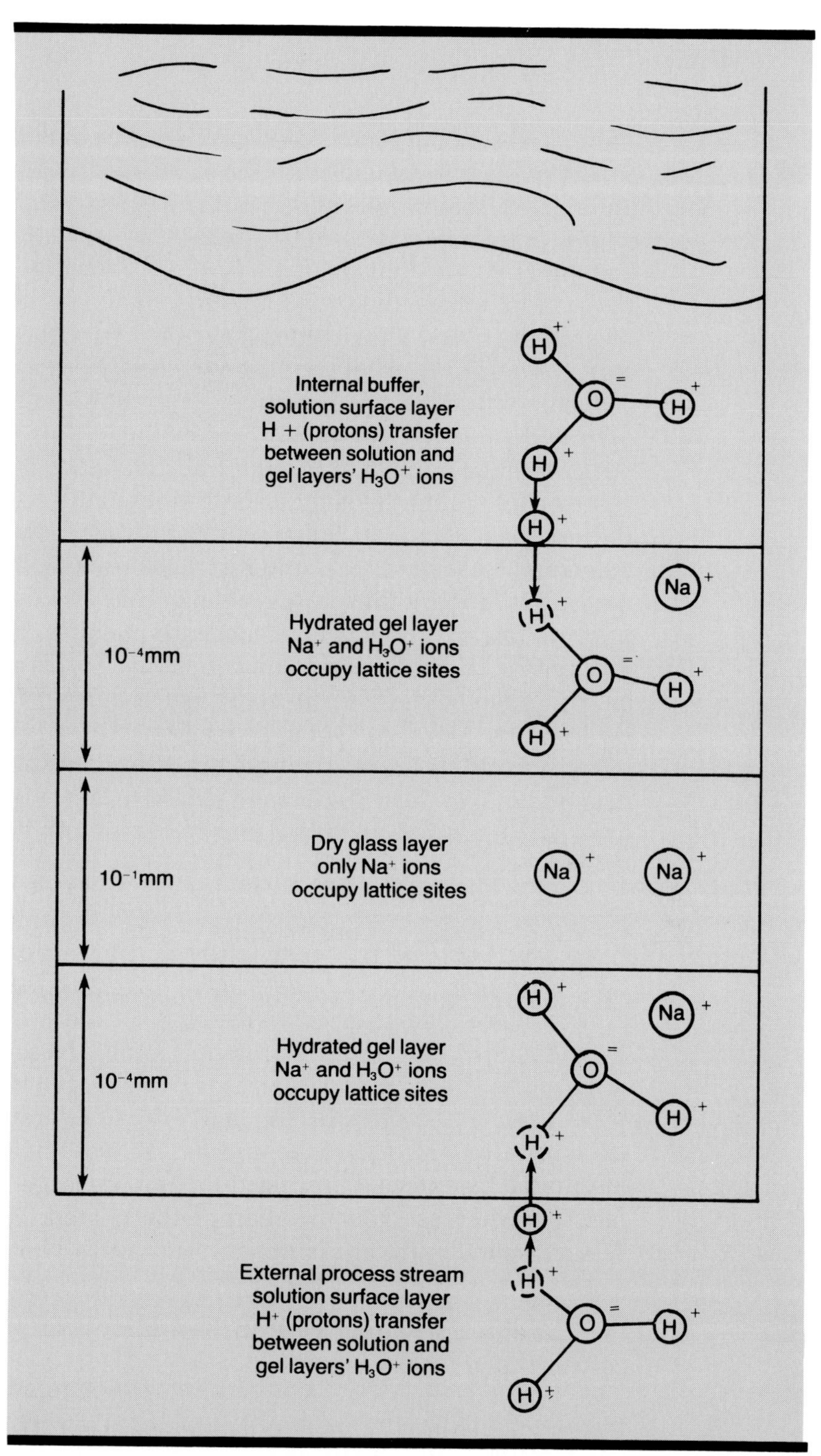

Fig. 4-2. A potential is developed at both the inner and outer glass surfaces of the measurement electrode by proton exchange with a hydrated gel layer.

direct submersion in agitated vessels. The accuracy is 0.1 pH up to 6 pH at the temperature rating.

The actual sensing of pH is accomplished by having a pH-sensitive glass in contact with the internal fill, a 7 pH buffer, and the external sample or stream. The pH-sensitive glass develops potentials per Eqs. (4-1a) and (4-2a), which are Nernst equations, by the hydrogen ion (proton) exchange between hydronium ions in the aqueous solutions and in the hydrated gel layer of the glass. The protons leave the hydronium ions in the aqueous solution, enter the sites vacated by positive alkali ions such as sodium, and recombine with hydronium ions in the hydrated gel layer. The potential developed is proportional to the difference in logarithms of the activity of hydronium ions in solution and in the gel layer on both sides of the glass membrane. If the gel layers have an equal number of sites for proton exchange, the constants Kg1 and Kg2 will be equal. If all the original sodium ions at these sites in the gel are also replaced by protons, the activities ag1 and ag2 are equal (Ref. 3). If these glass gel constants and activities are equal, Eqs. (4-1b) and (4-2b) can be combined to yield Eq. (4-3). By use of the definition of pH per Eq. (2-6), the logarithms of hydrogen activity can be converted to pH which yields Eq. (4-4) where the difference in potentials is proportional to the difference in pH. Also, since the internal fill has a hydrogen activity that corresponds to 7 pH, the equation for the potential difference is simplified to that shown in Eq. (4-4). Examination of this equation for the pH measurement electrode yields the following conclusions:

Conclusions from Eq. (4-4) for pH Measurement Electrode

(1) The millivolt output of the electrode decreases as the pH increases.

(2) The millivolt output is zero at 7 pH.

(3) The millivolt output is positive below 7 pH and negative above 7 pH.

(4) The effect of temperature on the millivolt output approaches zero as the pH approaches 7.

(5) At 25 degrees C, the output changes by 59.16 millivolts per pH unit.

$$E_1 = K_{g1} + 0.1984*(T + 273.16)*\log \frac{a_1}{a_{g1}} \quad (4\text{-}1a)$$

$$E_1 = K_{g1} + 0.1984*(T + 273.16)*[\log(a_1) - \log(a_{g1})] \quad (4\text{-}1b)$$

$$E_2 = K_{g2} + 0.1984*(T + 273.16)*\log \frac{a_2}{a_{g2}} \quad (4\text{-}2a)$$

$$E_2 = K_{g2} + 0.1984*(T + 273.16)*[\log(a_2) - \log(a_{g2})] \quad (4\text{-}2b)$$

If $K_{g1} = K_{g2}$ and $a_{g1} = a_{g2}$, then

$$E_1 - E_2 = 0.1984*(T + 273.16)*[\log(a_1) - \log(a_2)] \quad (4\text{-}3a)$$

$$E_1 - E_2 = 0.1984*(T + 273.16)*(pH_2 - pH_1) \quad (4\text{-}3b)$$

$$E_1 - E_2 = 0.1984*(T + 273.16)*(7 - pH_1) \quad (4\text{-}4)$$

where:
a_1 = activity of hydrogen ion in external process fluid (normality)

a_2 = activity of hydrogen ion in internal fill fluid (normality)

a_{g1} = activity of hydrogen ion in outer gel surface layer (normality)

a_{g2} = activity of hydrogen ion in inner gel surface layer (normality)

E_1 = potential developed at external glass surface (millivolts)

E_2 = potential developed at internal glass surface (millivolts)

K_{g1} = constant for potential for outer gel surface layer (millivolts)

K_{g2} = constant for potential for inner gel surface layer (millivolts)

pH_1 = pH of external solution

pH_2 = pH of internal solution (typically 7 pH)

T = solution temperature (degrees C)

4-2. Electrode Error

While pH measurement techniques have improved in recent years, many pH control systems still fail due to excessive measurement error. pH control is difficult enough without having measurement problems. Enough knowledge has been gained to avoid most of the measurement errors, however this expertise has only partially been documented and is scattered among many engineers and technicians who have faced and solved individual problems. This section will summarize the current state of knowledge from diverse experience and present it within the framework of an electrical circuit in which errors are represented by extraneous potentials and resistances.

The accuracy attainable with pH electrodes in a laboratory environment under ideal conditions is impressive. The short-term repeatability for a standard set of electrodes under ideal conditions is about ± 0.01 millivolts. For a solution temperature of 25°C, Eq. (4-4) shows that the electrode potential changes by 59.16 mv/pH. Thus, a change of ± 0.01 millivolt corresponds to a change of ± 0.00017 pH. The difference in internal and external potential at 7 pH, which is called the asymmetry potential, changes as the electrode membranes age. The drift in asymmetry potential for a standard set of electrodes without any problems is about 0.001 millivolts per day or 0.000017 pH units per day (Ref. 4). These repeatability and drift errors are for ideal conditions where identical test solutions at identical temperatures are used and the solutions have sufficient conductivity and a pH that does not cause either an acidic or alkaline error for the measurement electrode.

pH measurement accuracy also depends on bias, span, and nonlinearity errors from a variety of sources. Even under the best of industrial conditions, pH electrode potential measurements are not more accurate than about ± 1 millivolt or ± 0.017 pH (Ref. 5). This error translates to a percent of reading error that is equal to ± 0.13% at 13 pH (100%*0.017pH/13pH) and equal to ± 1.7% at 1 pH (100%*0.017pH/1pH). The titration curve slope is frequently flat at low pH so that the pH measurement error for a low setpoint may be larger than the pH error due to reagent delivery error from influent or reagent disturbances (the measurement error is greater than the control error). Table 1-1, which lists the hydrogen ion concentration for 0 to 14 pH, shows that a 0.017 pH error represents a relatively large hydrogen ion concentration error at low pH and an incredibly small hydrogen ion concentration error at high pH.

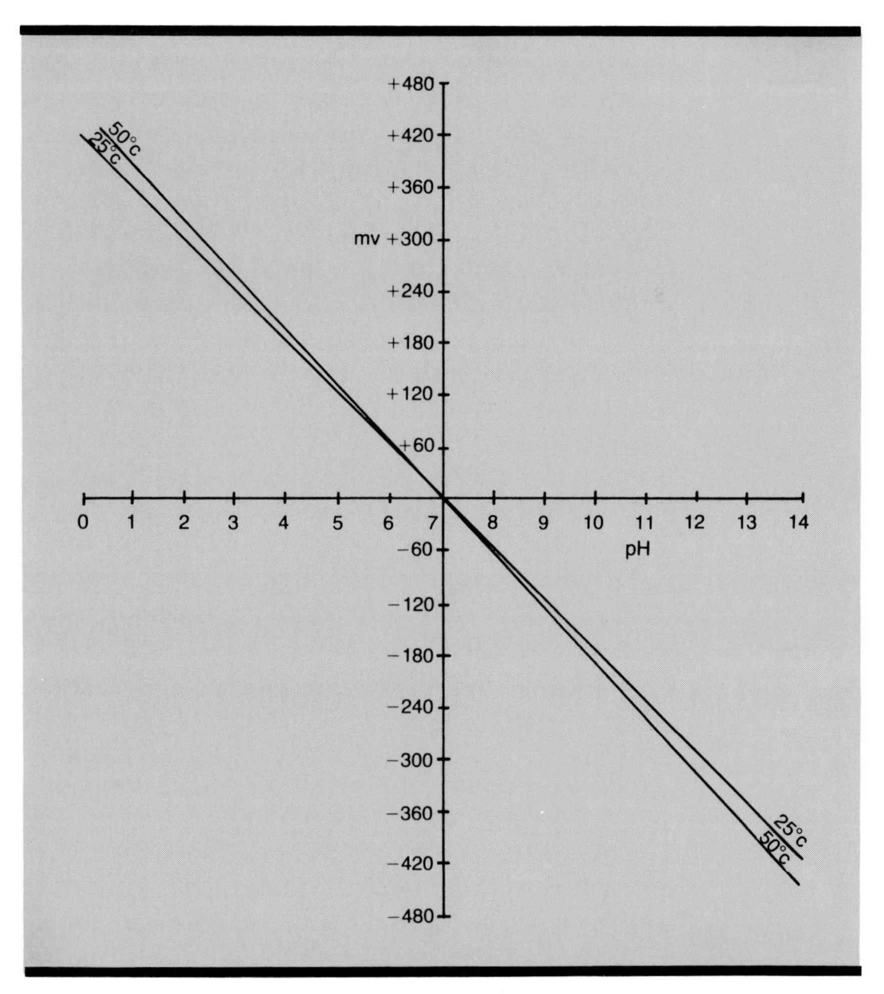

Fig. 4-3. The pH electrode error due to changes in solution temperature is small near 7 pH.

Equation (4-4) shows that the potential difference will change at any pH other than 7 if the solution temperature changes. The magnitude of the pH error depends upon both the magnitude of the temperature error and the deviation of the pH reading from 7 pH. The size of the error is usually small because change in the process temperature is small compared to the 273.16 in Eq. (4-4) used to convert to degrees Kelvin and because most pH control system setpoints are near 7 pH. Automatic temperature compensators are installed in most systems as a matter of practice. The widespread use of unnecessary automatic temperature compensators is not a problem because their complexity, cost, and failure rate are small. **Automatic temperature compensators do not correct for the change in solution pH with temperature.** An increase in solution temperature from 25 to 50°C, means that the change in potential with pH is now 64.12 mv/pH instead of 59.16 mv/pH. This change in the slope of the millivolt versus pH line is shown in Fig. 4-3. An automatic temperature compensator consists of a thermistor immersed in the solution in which resistance changes with temperature to change the pH meter amplifier gain in a direction equal but opposite to the change in slope. This resistance is located next to the pH meter span resistance in the feedback path of the F.E.T. amplifier as shown in Fig. 2-4. Equations (4-1b) and (4-2b) show that an asymmetry potential error from constant Kg1 not equal to constant Kg2 can be cancelled by adjusting the standardization potential in Eq. (4-5a) (Es = Kg2 − Kg1) without affecting the accuracy of the temperature compensator since these constants do not depend on temperature. However an asymmetry potential error from activity ag1 not equal to activity ag2 can only be accurately cancelled by adjusting the standardization potential in Eq. (4-5a) (Es = 0.1984*(273.16 + T)*[log(ag1)−log(ag2)]) at a given temperature since the potentials from these activities depend on temperature.

Temperature compensation errors also arise from nonidealities and dynamics. Figure 4-3 shows an intersection of the different slopes at a point that is called the isopotential point. A change in process fluid temperature has no effect at this point. Equation (4-4), which is for ideal conditions, shows that when the pH is 7, the temperature has no effect since it is multiplied by zero. An exception to this rule is the Pfaudler Glasteel electrode in which zero potential occurs at about 2 pH. The actual location of the isopotential point will be shifted due to various nonidealities. Also, the intersection is not actually a point but an area so that an error exists between the position of the real line and that corrected for by the temperature compensator (Ref. 6). The accuracy of slope duplication depends on the linearity of the resistance change of the thermistor. Also, the pH measurement electrode potential will normally change faster than the thermistor resistance for a temperature change so that a dynamic error is created. For a ramping change in solution temperature, the temperature compensation change has a time delay approximately equal to the time constant of the temperature sensor.

The manual temperature compensation adjustment consists of a potentiometer to duplicate the resistance and slope change performed by the thermistor of the automatic temperature compensator. A common mistake is to try to adjust the manual temperature compensation at 7 pH where it has no effect.

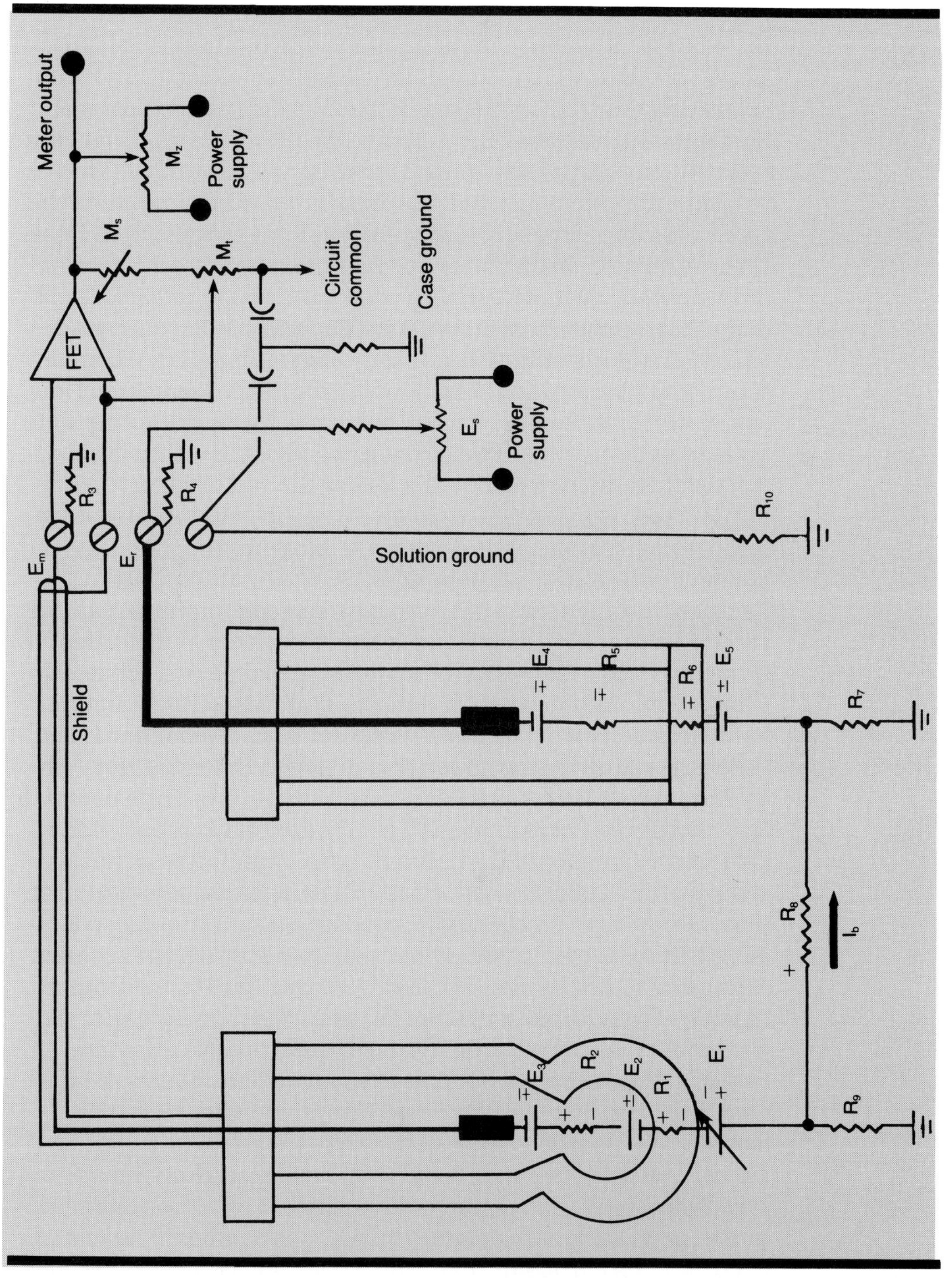

Fig. 4-4. The pH electrode and meter system has many potentials and resistances that create errors and many calibration adjustments to correct for these errors.

The calibration adjustments in addition to manual temperature compensation are shown in Fig. 4-4, Eqs. (4-5a) through (4-5c), and Table 4-1. By understanding the effect of each adjustment on the millivolt versus pH line, the proper adjustment can be selected for a given type of error. Note that there is a meter zero (bias) and span adjustment and a transmitter output zero and span adjustment. The meter adjustments are used to insure that the meter indication is correct over the 0 to 14 pH range and the transmitter output adjustments are used to insure the voltage or current output matches the selected output range for the control room indicator, recorder, or controller. For example, the output zero might be adjusted to provide a 4 ma output for 2 pH and the output span adjusted to provide a 20 ma output at 12pH. Both the manual temperature compensation and meter span adjustments change the slope of the line. The manual temperature compensation adjustment should be used for the purpose given in its name. The manual temperature adjustment dial has graduations in degrees so that it can be visually set once the solution temperature is known. The meter span should be used to correct for the other slope errors in Table 4-4, which are caused by measurement bulb coating, abrasion, etching, and aging. Of course, recalibration should not be used as a coverup for a recurring problem. The source of and remedy for the problem should be sought. When the standardization adjustment is used to correct for horizontal shifts of the isopotential point caused by measurement electrode abrasion, dehydration, etching, and contamination, an error is created in the temperature compensation. When the standardization adjustment is used to correct for vertical shifts of the isopotential point caused by a low temperature of the measurement electrode, fill contamination and bulb coating of the reference electrode, and changes in solution resistance and composition (liquid-junction potential), a temperature compensation error is not created if the standardization potential exactly offsets the potential of the error in Eq. (4-5a). Figure 4-5a shows the family of lines for a horizontal leftward shift of the isopotential point without temperature compensation and Fig. 4-5b shows the same family but with temperature compensation based on the original isopotential point at 7 pH. Figure 4-6a shows the family of lines for a vertical downward shift of the isopotential point without temperature compensation and Fig. 4-6b shows the same family but with temperature compensation based on the original isopotential point at 7 pH. If the temperature compensation was perfect in Figs. 4-5b and 4-6b, the family of lines for all temperatures would merge into a single line. Both isopotential shifts cause an upscale shift of the pH reading since

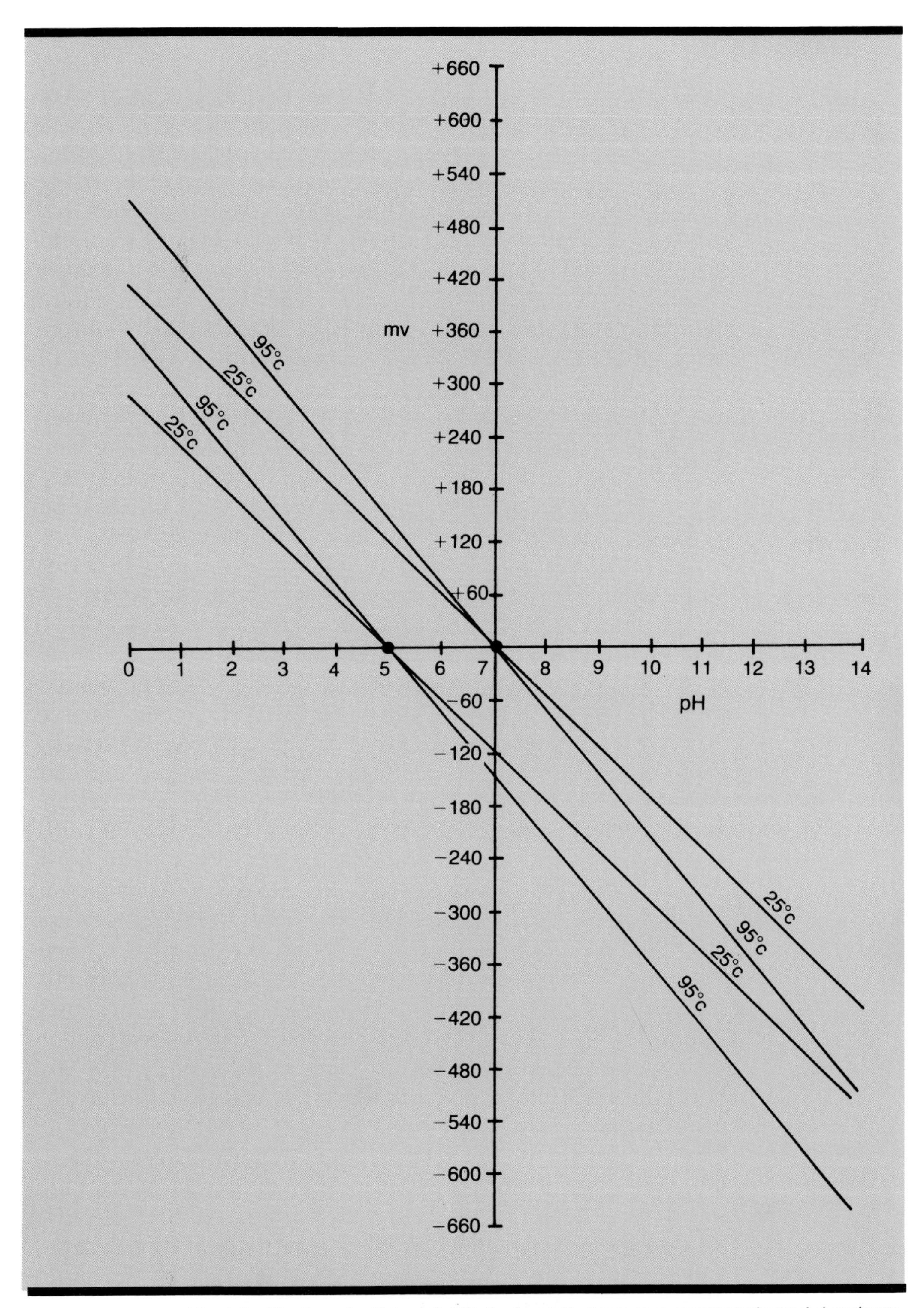

Fig. 4-5a. The isopotential point shifts horizontally due to measurement electrode breakage, abrasion, dehydration, etching, and contamination.

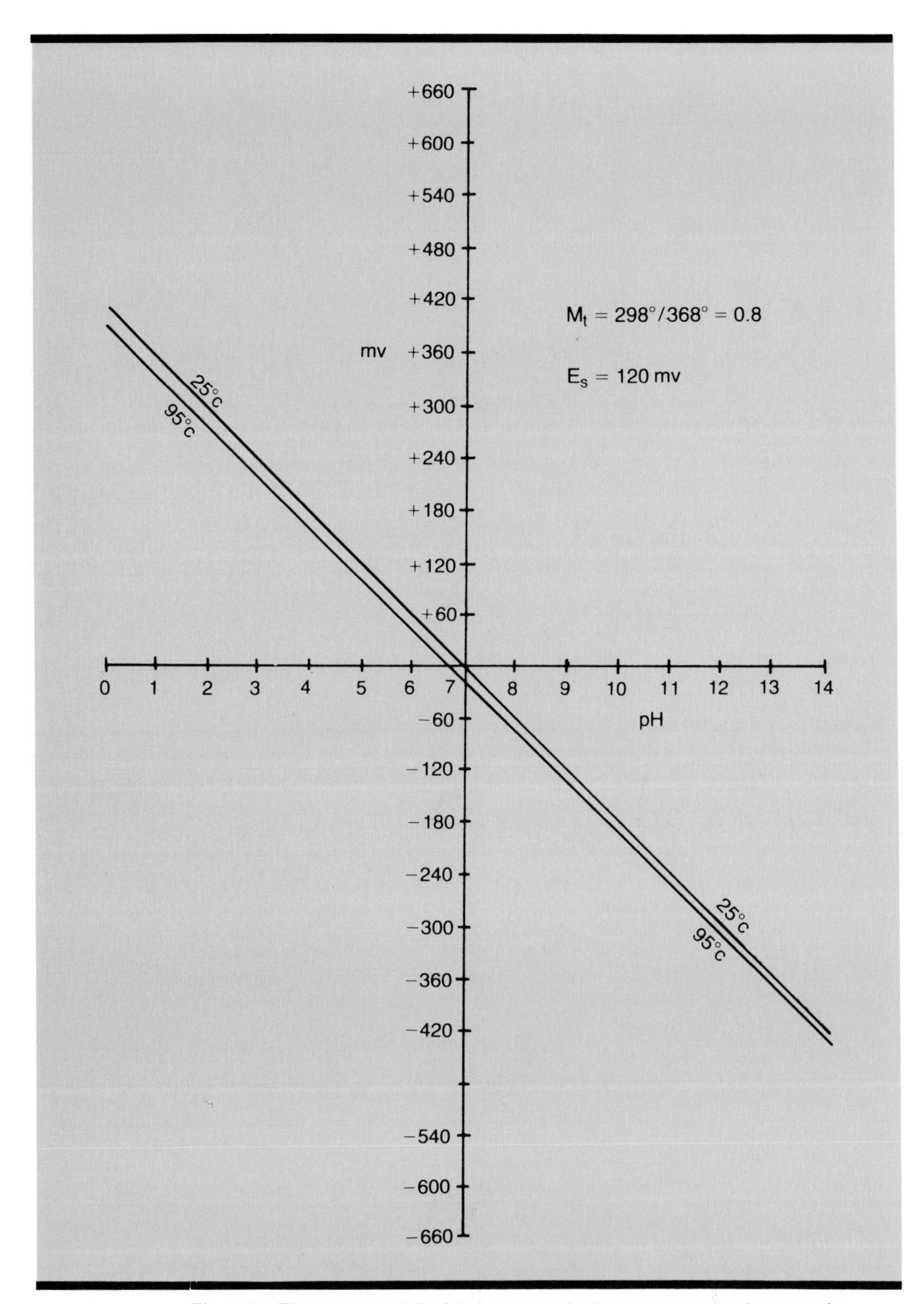

Fig. 4-5b. The horizontal shift of the isopotential point creates an error in automatic temperature compensation that cannot be eliminated by the standardization adjustment.

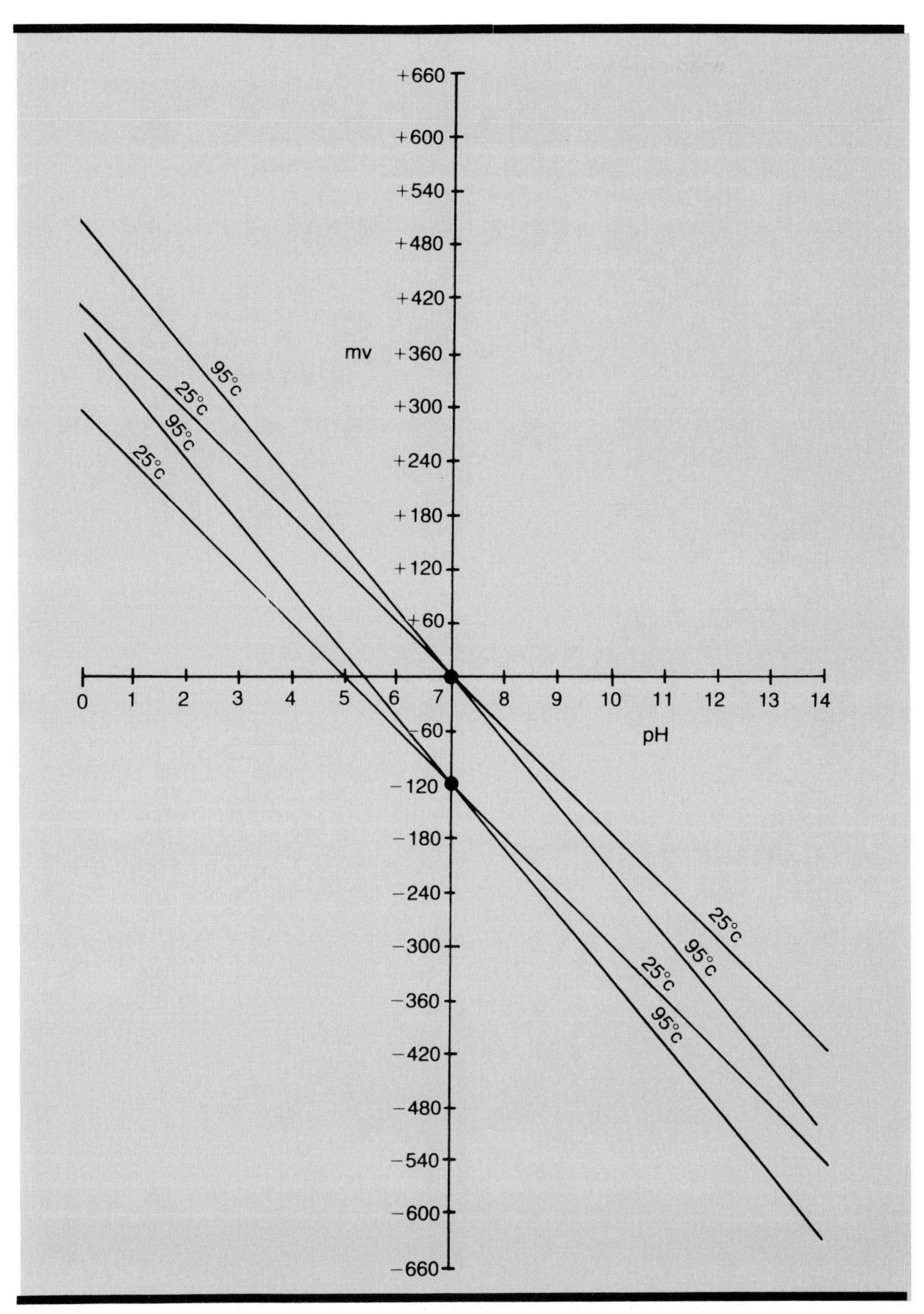

Fig. 4-6a. The isopotential point shifts vertically due to measurement reference electrode contamination and coating, high solution resistance and changes in composition, and low measurement electrode temperature.

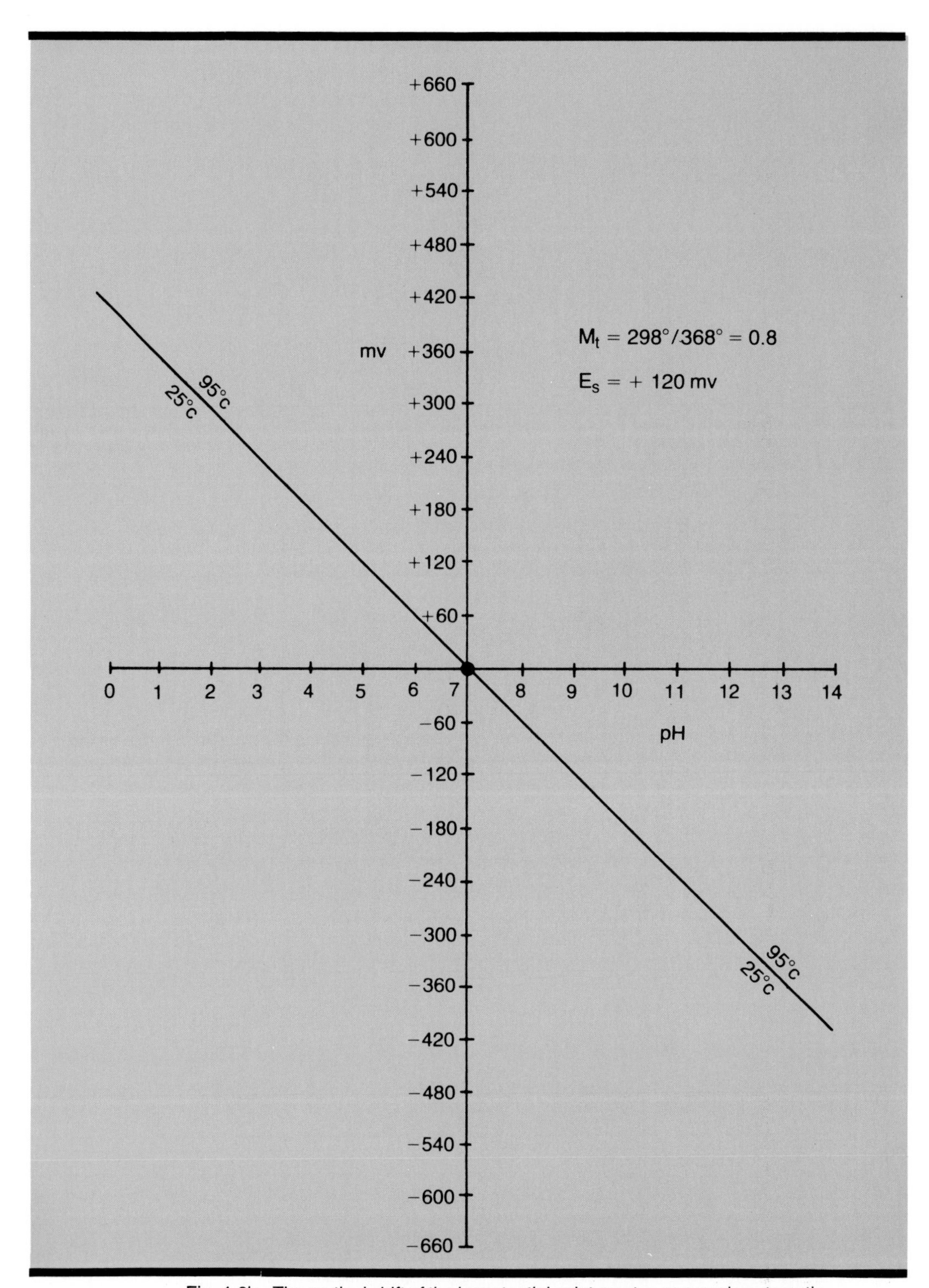

Fig. 4-6b. The vertical shift of the isopotential point creates an error in automatic temperature compensation that can be eliminated by the standardization adjustment.

the millivolt versus pH line has a negative slope. The meter zero is normally used to compensate for errors in the pH meter electronics or meter pointer movement. This adjustment is typically done with the electrode input terminals shorted to provide a zero millivolt or 7 pH signal that is independent of any electrode problems.

Type of Calibration Adjustment	Effect on Line in Fig. 4-3
Manual temperature compensator M_t	Slope increased or decreased
Meter span M_S	Slope increased or decreased
Meter zero M_Z	Isopotential point shifted vertically
Standardization E_S	Isopotential point shifted horizontally
Output span O_S	Output span increased or decreased
Output zero O_Z	Output zero increased or decreased

Table 4-1. The four meter adjustments change the slope or shift the isopotential point of the millivolt versus pH line and the two output adjustments set the pH range transmitted to the control room.

The potential of interest in pH measurement is the difference between the potential developed at the outer and inner glass surfaces of the measurement electrode as defined by Eq. (4-4). Any other potential represents an error. Figure 4-4 shows the physical location of each potential and Eq. (4-5a) shows that the effect of these potentials is additive. Whereas errors due to changes in parameters in Eqs. (4-1) and (4-2) result in either horizontal shifts of the isopotential point or changes in the slope of the line, all extraneous potentials in Eq. (4-5a) result in a vertical shift of the isopotential point.

$$E_i = E_1 - E_2 - E_3 + E_4 + E_5 - I_i * (R_1 + R_2 + R_5 + R_6 + R_8) + E_s \quad (4\text{-}5a)$$

$$E_o = M_t * M_s * E_i + M_z \quad (4\text{-}5b)$$

$$I_o = O_s * E_o + O_z \quad (4\text{-}5c)$$

where:

E_1 = potential developed at external glass surface (millivolts)

E_2 = potential developed at internal glass surface (millivolts)

E_3 = half-cell potential of the measurement electrode (millivolts)

E_4 = half-cell potential of the reference electrode (millivolts)

E_5 = liquid junction potential of the reference electrode (millivolts)

E_i = transmitter input voltage (millivolts)

E_o = meter output voltage (millivolts)

E_s = electrode standardization potential (millivolts)

M_z = meter zero adjustment (bias)

M_s = meter span adjustment (gain)

M_t = meter temperature compensation adjustment (gain)

O_z = transmiter output zero adjustment (bias)

O_s = transmitter output span adjustment (gain)

I_i = input leakage current of the meter amplifier (milliamps)

I_o = transmitter current output (milliamps)

R_1 = measurement electrode glass resistance (ohms)

R_2 = measurement electrode internal fill resistance (ohms)

R_5 = reference electrode internal fill resistance (ohms)

R_6 = reference electrode liquid junction resistance (ohms)

R_8 = solution resistance between measurement and reference electrodes (ohms)

The measurement and reference electrode half-cell potentials in Eq. (4-5a), which are due to an electrochemical reaction between the internal electrodes and fill, are of opposite sign and should ideally be equal so that their sum is zero. However the half-cell potentials depend upon the electrode type (silver chloride or calomel), the internal fill concentration, and the electrode temperature. If the electrode type and fill are identical, then the change in half-cell potential with temperature will cancel out unless a temperature gradient exists between the reference and measurement electrode locations. The half-cell potential for a saturated calomel electrode ranges from 234 mv to 40° to 260 mv at 0°C and for a saturated silver chloride electrode ranges from 193 mv at 40° to 237 mv at 0°C (Ref. 7).

The liquid junction potential is located at the tip of the reference electrode. If the ions in the internal fill and the external solution have different mobilities (speeds attained from an electrical force), ion concentration and charge gradients will develop. For example, the hydrogen ions have a much greater mobility than the chloride ions of a concentrated hydrochloric acid solution. Since ions diffuse to regions of low concentration, the hydrogen and chloride ions will both diffuse to the internal reference electrode fill. If the two ions had identical mobilities, the same number of positive hydrogen charges and negative chloride ion charge would arrive at the electrode tip at any given time. However, the hydrogen ion is faster so that more positive charges arrive and accumulate at the tip. The charge difference increases until it is large enough to suppress hydrogen ion movement in excess of chloride ion movement to the tip. An equilibrium is then established with a potential difference that has the same sign as the measurement electrode's outer potential. The isopotential point is shifted up and the pH measurement goes downscale. Table 4-2 shows that the liquid potential size is greatest for concentrated strong acids and that the liquid potential sign is positive for acids and dilute bases and negative for concentrated bases. The liquid potential can be reduced by placing the reference electrode in a glass body with an open tip and filled with a salt solution to form what is called a "salt bridge" between the reference electrode and the process solution. The salt selected must have positive and negative ions of nearly equal mobility and a high concentration, such as saturated potassium chloride. The salt ions must not react with or contaminate either the electrode fill or process fluid. The salt solution pressure must be maintained greater than the process fluid by external gas pressurization if necessary. "Salt bridge" electrode assemblies are also called "double junction" or "flowing junction" electrodes (Ref. 8).

External Fluid Type	Concentration Molarity	Potential (MV)
Hydrochloric acid	1.0	14.1
Hydrochloric acid	0.1	4.6
Hydrochloric acid	0.01	3.0
Potassium chloride	0.1	1.8
Sodium hydroxide	1.0	−8.6
Sodium hydroxide	0.1	−0.4
Sodium hydroxide	0.01	2.3
Potassium hydroxide	1.0	−6.9
Potassium hydroxide	0.1	−0.1

Table 4-2. The liquid junction potential size is greatest for concentrated strong acids (above potentials are for a reference electrode with an internal fill saturated with potassium chloride).

The resistances in Eq. (4-5a) are all relatively large. Fortunately, the input leakage current that flows through these resistances and the amplifier input is extremely small (about 1 picoamp or one trillionth of a milliamp) (Ref. 9). The current flows from the positive measurement electrode terminal to the negative reference electrode terminal so that the sign of the potential drop is negative compared to the measurement electrode's outer potential and consequently causes the isopotential point to shift down and the pH measurement to go upscale.

The measurement electrode glass bulb resistance is normally the largest resistance. It is about one tenth of a trillion ohms so that the potential drop is about 0.1 millivolts. However the glass resistance will dramatically increase as the process fluid temperature decreases. Figure 4-7 shows that the resistance of Corning 015 glass increases from about 200 megohms (200 million ohms) at 25°C to about 1000 megohms at 10°C, at which the resistance is offscale (Ref. 10). Thus, even though nearly all electrodes are listed as having a temperature range that starts at 0°C, the actual resistance at the minimum process fluid temperature should be checked. Electrodes specifically designed for a low alkaline error or high temperature have a higher than normal resistance and should therefore not be used at low temperatures. Electrodes whose gel layer has dehydrated will have an abnormally high resistance.

The resistances of the internal electrode fills and external process fluid are normally only a few thousand ohms so that the potential drop due to leakage current flow is negligible. However, high purity water (i.e., distilled water, deionized water, or steam

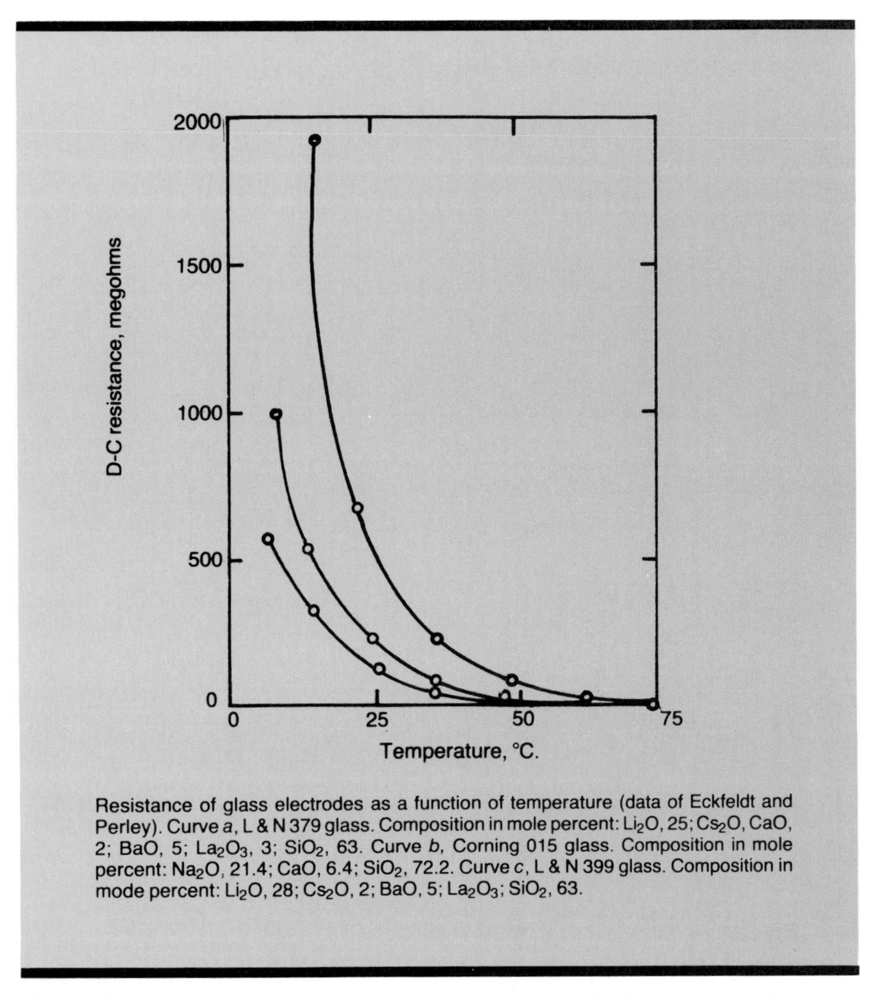

Resistance of glass electrodes as a function of temperature (data of Eckfeldt and Perley). Curve *a*, L & N 379 glass. Composition in mole percent: Li_2O, 25; Cs_2O, CaO, 2; BaO, 5; La_2O_3, 3; SiO_2, 63. Curve *b*, Corning 015 glass. Composition in mole percent: Na_2O, 21.4; CaO, 6.4; SiO_2, 72.2. Curve *c*, L & N 399 glass. Composition in mode percent: Li_2O, 28; Cs_2O, 2; BaO, 5; La_2O_3; SiO_2, 63.

Fig. 4-7. The measurement electrode glass resistance dramatically increases and causes a large positive pH error near 0°C. (From R. G. Bates, *Determination of pH: Theory and Practice*, New York, John Wiley & Sons, Inc., 1964, p. 303)

condensate) and nonaqueous solutions have extremely high resistances. Equation 4-6 shows that the resistance is inversely proportional to the conductivity of the solution and is proportional to the distance between the electrodes.

The distance between the measurement and reference electrode should be minimized for low conductivity process fluids. While a combination electrode reduces this distance by the electrodes sharing the same internal fill, the chloride ions may precipitate upon contact with high purity water, create a liquid junction potential, and coat both bulbs. A high flow junction on the reference electrode may add enough ions by leakage of the internal fill to increase the conductivity of the process fluid (the concentration required is extremely low). For details on how to estimate the electrical conductivity of various aqueous and non-aqueous fluids at different concentrations, the reader is directed to Ref. 11. The special considerations in electrode assembly design required for these measurements are described in Refs. 12 and 13.

$$R = \frac{L}{A*C} * 10^6 \quad (4\text{-}6)$$

where:
A = cross sectional area of electrode surface (cm^2)
C = conductivity of fluid (micromhos/cm)
L = distance between electrodes (cm)
R = resistance of fluid (ohms)

Table 4-3 summarizes the different conditions and symptoms for many of the more common sources of errors. Just the sheer quantity of sources shows you that pH measurement accuracy can easily deteriorate and that pH measurement troubleshooting can be difficult. By identifying these sources, installations and operating conditions that are prone to cause these problems can be avoided. By identifying the symptoms, the problems that do develop can be remedied.

Chunks of solids impinging on the electrode, chemical attack of glass by hydrofluoric acid, or just old age can lead to *electrode breakage*. The measurement electrode gel layer wears away and penetrates deeper into the glass with time. Even under the best of conditions, the useful life of a measurement electrode varies from nine months to two years. If the measurement electrode bulb breaks, process fluid will be in contact with both the inside and outside pH sensitive glass layers. The hydrogen activity, and hence the potential, will be about equal so that the difference in potentials is nearly zero, which corresponds to a constant 7 pH reading. However, the measurement electrode glass resistance is bypassed (paralleled) by a much smaller process fluid resistance so that R1 decreases and the measured potential increases per Eq. (4-5a). The increase in measured potential corresponds to a decrease in pH reading per Eq. 4-4. The result is a shift of the isopotential point vertically upward and a counterclockwise rotation of the millivolt versus pH line to a horizontal position as shown in Fig. 4-8. To determine what the pH reading would be for this position of the line, the original position of the line used

Source of Error	Electrical Symptom	Response Symptom	Effect on Line in Fig. 4-3
Measurement electrode:			
Bulb broken	$E_1 = E_2$ and $R_1 \downarrow$	No response (4 to 6 pH)	Horizontal line
Fill contamination	$E_1 = E_2$ and $R_1 \downarrow$	No response (4 to 6 pH)	Horizontal line
Bulb abrasion	$\lvert \Delta E_1/\Delta pH \rvert \downarrow$ and $E_1 \downarrow$	Slow, erratic, shortened span and upscale pH	Slope magnitude decreased and isopotential point shifted left
Bulb dehydration	$E_1 \downarrow$, $R_1 \uparrow$ and $\lvert \Delta E_1/\Delta pH \rvert \downarrow$	Slow, erratic, shortened span, and upscale pH	Isopotential point shifted left and down and slope magnitude less
Bulb etching	$\lvert \Delta E_1/\Delta pH \rvert \downarrow$ and $E_1 \downarrow$	Slow, erratic, shortened span and upscale pH	Slope magnitude decreased and Isopotential point shifted left
Partial bulb coating	$\lvert \Delta E_1/\Delta pH \rvert \downarrow$	Very slow	Slope decreased
Complete bulb coating	E_1 fixed	No response	Horizontal line
Low temperature	$R_1 \uparrow$	pH increases as temperature decreases	Isopotential point shifted down
Reference electrode:			
Bulb broken	$E_4 \downarrow$ or $\uparrow$	Drift upscale or downscale	Isopotential point shifted down or up
Fill contamination	$E_4 \downarrow$ or $\uparrow$	Drift upscale or downscale	Isopotential point shifted down or up
Partial bulb coating	$E_5 \downarrow$ and $R_6 \uparrow$	Drift upscale (typically)	Isopotential point shifted down
Complete bulb coating	Open circuit	Offscale up or down, depending on meter type	Horizontal line
Thermistor:			
Open circuit	$\lvert \Delta E_1/\Delta pH \rvert \downarrow$	Shortened span	Slope magnitude decreased
Shorted circuit	$\lvert \Delta E_1/\Delta pH \rvert \uparrow$	Lengthened span	Slope magnitude increased

Table 4-3. Due to numerous possible errors, proper diagnosis requires knowledge of the electrical symptoms, response symptoms, and the effects on the millivolt versus pH line.

Source of Error	Electrical Symptom	Response Symptom	Effect on Line in Fig. 4-3
Solution:			
Acidic solvent (no water)	$R_8 \uparrow E_5 \uparrow E_1 \uparrow$	pH offscale downwards	Isopotential point shifted down and increased proton activity
Basic solvent (no water)	$R_8 \uparrow E_5 \downarrow E_1 \downarrow$	pH offscale upwards	Isopotential point shifted down and decreased proton activity
Alcohol solvent (no water)	$R_8 \uparrow$ and $E_5 \downarrow$	pH upscale	Isopotential point shifted down
Hydrocarbon solvent (no water)	$R_8 \uparrow$ and $E_5 \downarrow$	pH upscale and decreased lower and increased upper limits	Isopotential point shifted down and increased line length
Pure water	$R_8 \uparrow$	pH upscale and erratic	Isopotential point shifted down
Composition changes	$E_5 \downarrow$ or $\uparrow$	Drift upscale or downscale	Isopotential point shifted down or up
Gas bubbles	$E_1 \downarrow$ and $R_8 \uparrow$	Upscale noise	Isopotential point shifted left and down randomly
Low pH (acid error)	$\lvert \Delta E_1/\Delta pH \rvert \downarrow$	Shortened span at low pH end	Bends over at low pH end
High pH (alkalinity or sodium ion error)	$\lvert \Delta E_1/\Delta pH \rvert \downarrow$	Shortened span at high pH end	Bends over at high pH end
Low temperature	$\lvert \Delta E_1 /\Delta pH \rvert \downarrow$	Shortened span	Slope magnitude decreased
High temperature	$\lvert \Delta E_1 /\Delta pH \rvert \uparrow$	Lengthened span	Slope magnitude increased
Terminals:			
Short form M to R	$E_i = 0$	Fixed at 7 pH	Horizontal line on abscissa
Broken electrode wire	$E_i = 0$	Fixed at 7 pH	Horizontal line on abscissa
Short form M to ground	$E_i = 0$ & $R_3 = 0$	Fixed at 7 pH	Horizontal line on abscissa
Moisture on M	$R_3 \downarrow$	Stays near 7 pH	Slope magnitude decreased
Moisture on R	$R_4 \downarrow$	Upscale pH	Isopotential point shifts downscale

Table 4-3 continued

in the calibration must be known. For example, if the calibration is represented by the line position shown in Fig. 4-3, the horizontal line in Fig. 4-8 at 120 millivolts would give an indication of 5 pH. Field experience shows that a broken measurement electrode causes a constant pH reading between 4 and 6 pH (Ref. 14). If the reference electrode breaks, the internal fill is contaminated with the process fluid. The reference electrode half-cell potential will change and shift the isopotential point vertically upward or downward and the pH span will accordingly be biased in the opposite direction. Unlike the measurement electrode, the reference electrode fill can be contaminated without breakage due to its liquid junction. Pulsating pressures or high ion concentrations can force ions from the process fluid back through the liquid junction into the fill.

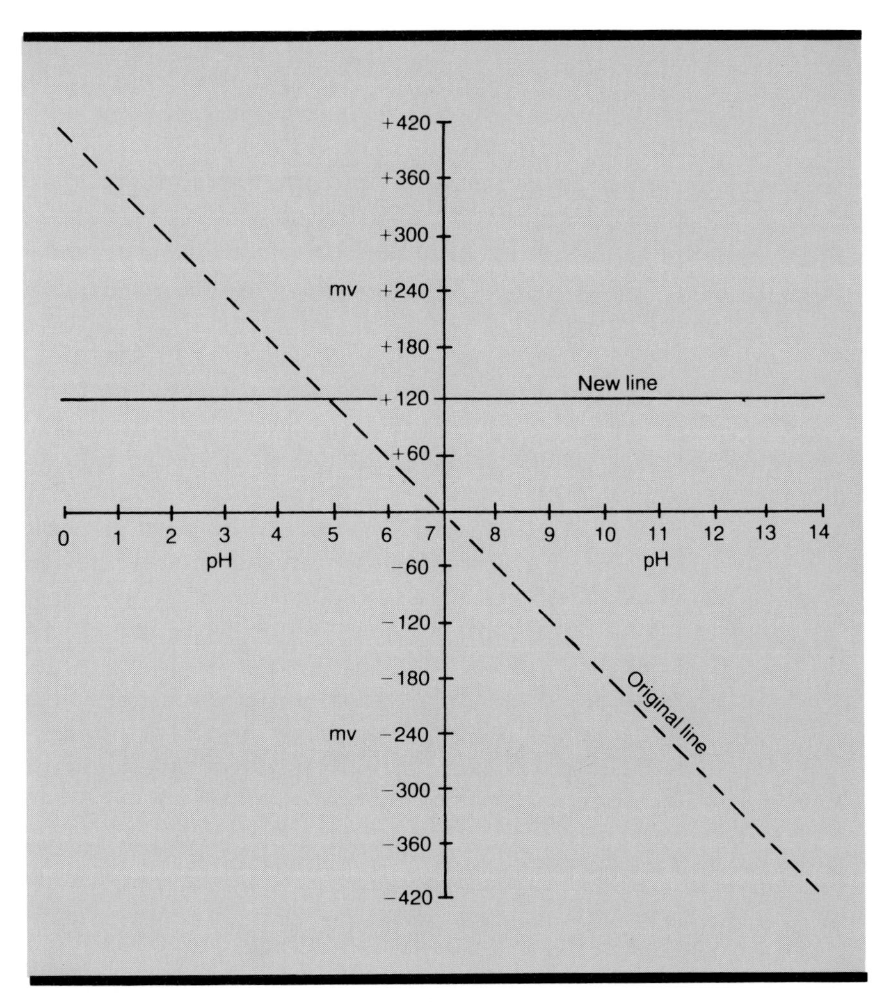

Fig. 4-8. Measurement electrode breakage results in a constant millivoltage corresponding to 5 to 6 pH reading.

Slurries can cause bulb *abrasion*, nonaqueous fluids or air pockets can cause *dehydration*, and hydrofluoric acid can cause *etching*. The response symptoms for bulb abrasion, dehydration, and etching are the same in Table 4-3. In each case, the outer glass layer is less sensitive to hydrogen activity, or—in other words—a given activity will be measured as a lower activity which corresponds to a higher pH (pH is the inverse logarithm of activity per Eq. 2-6). The response span is shortened and offset upscale. If the transmitter is zeroed when immersed in a low pH buffer, the pH reading will be significantly low when immersed in a high pH buffer. The outer glass voltage for a given activity and the change in voltage for a change in activity will be less. The isopotential point is shifted left and the millivolt versus pH line slope is decreased by rotation of the line counterclockwise toward the horizontal position. The increase in the glass bulb resistance due to dehydration will cause the isopotential point also to shift vertically downward, which aggravates the upscale shift of the pH span. The response may become noticeably slower and erratic due to the reduced and irregular area for proton exchange between the fluid and the outer gel layer.

Oils, tars, gums, polymers, and particles can cause *coating* of measurement and reference electrodes. A partial coating of the measurement electrode forms a layer that slows down the diffusion of the hydrogen ions from the process fluid to the glass surface. The result is a very slow measurement response. The measurement time constant can increase from a few seconds to several minutes. The control loop period will dramatically increase. Equation (4-7) can be used to estimate the loop period for a large measurement time constant if the controller is not retuned. For a mixer turnover time of one minute and a measurement time constant that increases from two seconds to nine minutes, the loop period will increase from about four minutes to about 16 minutes (Ref. 15). A complete coating of the measurement electrode forms a barrier that stops all diffusion of the hydrogen ions from the process fluid to the glass surface. The result is a constant pH reading whose value depends upon the hydrogen activity of the coating. Since the hydrogen activity of most coatings is low, a constant upscale pH reading is common. This corresponds to a horizontal millivolt versus pH line at a negative millivolt level. A partial coating of the reference electrode will change the liquid junction potential and the resistance at the electrode tip. Usually the junction potential decreases and the resistance increases so that the isopotential point is shifted down and the pH span is shifted upscale. A complete nonconductive coating can cause an open circuit between the reference and measurement electrodes. The transmitter output will go off scale up or down, depending on the pH meter manufacturer and model number. The more typical failure mode is upscale with the pH meter output (transmitter input) approaching a minus one volt DC.

for $TC_m \geq 0.05*TD$:

$$T_c = 4* \left[1 + \sqrt{\frac{TC_m}{TD}} \right] * TD \tag{4-7}$$

where:
T_c = loop period if controller not retuned (minutes)

TC_m = measurement time constant (minutes)

TD = loop dead time (mixer turnover time) (minutes)

Nonaqueous solutions have no water. Instead of water, the solvent can be an acid, base, alcohol, or hydrocarbon. An *acid solvent* acts as a proton donor so that the proton activity is increased and the pH scale is shifted down. For example, the pH scale for an acetic acid solvent is −6 to −1 pH and for a formic acid solvent is −9 to −2 pH. A *base solvent* acts as a proton acceptor so that the proton activity is decreased and the pH scale is shifted up. For example, the pH scale for an ammonia solvent is 16 to 49 pH. An *alcohol solvent* acts both as a proton donor and acceptor like water so that changes in proton activity are moderated. The pH scale occupies about the same region as water but may extend further upscale and downscale. For example, the pH scale for an ethanol solvent is −4 to +16 pH. A *hydrocarbon solvent* acts neither as a proton donor nor an acceptor so that the solvent is passive to changes in proton activity. Consequently, the pH scale upper and lower limits are usually spread further apart than those for water. For example, the pH scale for an acetone solvent is −5 to 20 pH. Nearly all nonaqueous solvents have a higher resistance than water. So the isopotential point is shifted downscale and the pH span upscale. This shift is counteracted by the more positive liquid junction potential for acid solvents but is accentuated by the more negative liquid junction potential for base solvents. It is important to realize that dehydration, with all its attendant symptoms and errors, will also occur for all nonaqueous solutions unless the measurement electrode is periodically rinsed with water to replenish the hydronium ions in the gel layer. The start of dehydration is marked by a slowing down of the response (an increase in the loop period) and a drift of the measurement upscale. In a control loop, the measurement drift would not be noticeable until a base reagent valve had drifted all the way open or a acid reagent valve had drifted all the way closed. Thus, the output of the controller should be monitored for the early detection of pH measurement drift.

Pure water solutions have a high solution resistance like nonaqueous solutions. The addition of a small concentration of ions changes both the conductivity and pH tremendously, however the resulting in an erratic measurement besides the upscale shift of the pH span. Most pure water streams will absorb enough carbon dioxide upon exposure to air to exhibit a buffering effect below 7 pH.

Changes in *solution composition* will cause changes in the liquid junction potential of the reference electrode. If an acid concentration increases, the liquid junction potential increases and shifts the isopotential point up and pH span downscale. If a base concentration increases, the liquid junction potential decreases or becomes more negative and shifts the isopotential point down and the pH span upscale.

Gas bubbles can result from gas entrainment from excessive agitation, gas evolution from a reaction, or a gas reagent. The gas bubbles cause a high resistance and reduced hydrogen activity at the measurement electrode upon impact. The result is an intermittent shift of the isopotential point to the left and down which creates upscale pH noise. The standardization adjustment is used to center the noise band about the proper pH reading so that the short term plus and minus excursions about setpoint tend to cancel out. The transmitter's current output is either filtered or the controller's proportional band is increased (gain is decreased) to prevent the reagent valve from reacting to the noise. The standardization adjustment reduces the offset of the true pH reading from the setpoint. For details on the sequence of the calibration adjustments, the reader is directed to Unit 11 on pH control system checkout.

Alkalinity or *sodium ion error* is caused by alkali ions such as sodium ions penetrating the measurement electrode silicon oxygen network and creating a potential error at the outer electrode glass surface. The result is an indicated pH less than the true pH. The high end of the millivolt versus pH line bends over toward a horizontal position. The error is greatest for glasses with alkali ions of equal or larger radius than that of the alkali ions in solution. It is frequently called a sodium ion error because sodium ions are the most frequent cause due to the popularity of sodium hydroxide as a reagent and due to the penetrating capability of the relatively small radius of the sodium ion. The alkalinity error of Corning 015 glass for lithium and potassium ions is about one half and one fifth respectively of that for sodium ions due to their larger size. The alkalinity error also increases with temperature. For example, the alkalinity error of Corning 015 glass starts at 9.5 pH at 30°C, 8.0 pH at 50°C, and between 5 and 6 pH at 60°C (Ref. 16). The heat of neutralization from sodium hydroxide can cause a significant unsuspected alkalinity error for a 7 pH setpoint. Glass formulations have been made that reduce the alkalinity error but with some sacrifice as to nonlinearity over the 0 to 14 pH scale. Figure 4-9 shows that the correction pH increases exponentially with pH above some threshold level. Note that the solution pH is equal to the measured pH plus the correction pH and that the pH error is equal to the negative of the correction pH. For more information on the alkalinity error, the reader is directed to Ref. 16.

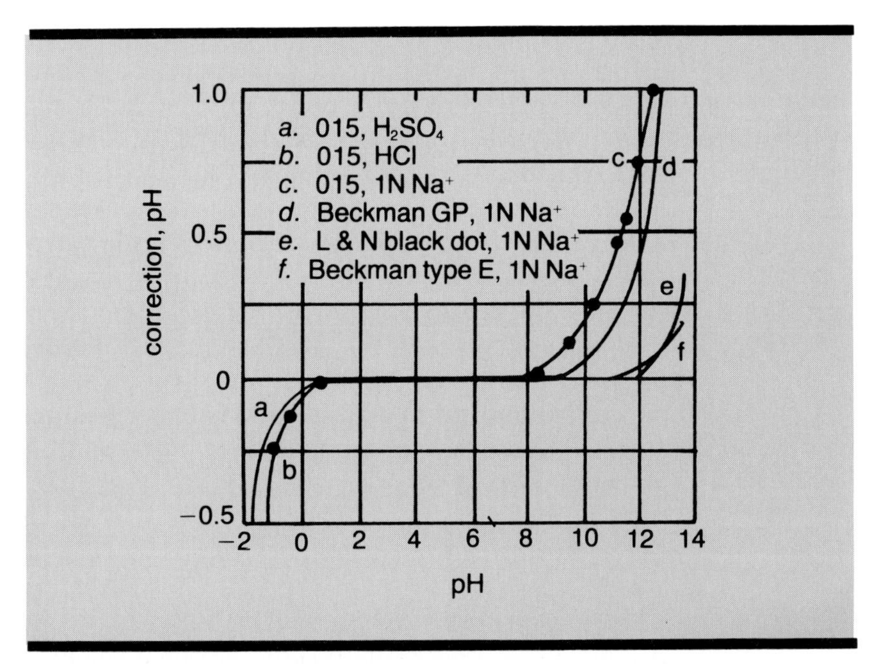

Fig. 4-9. The alkalinity error at the upper end and the acid error at the lower end of the pH scale depends on the type of measurement electrode glass. (From R. G. Bates, *Determination of pH: Theory and Practice,* New York, John Wiley & Sons, Inc., 1964, p. 316)

The cause of the *acid error* at the low end of the pH scale is not as well understood. The onset of the acid error is marked by a reduction in thickness of the outer hydrated gel layer of the measurement electrode. The result is an indicated pH that is greater than the true pH. The low end of the millivolt versus pH line bends over toward a horizontal position. The sign of the pH error is opposite of that for the alkalinity error. The error is independent of temperature but increases with time. Thus, significant upscale pH measurement drift at the low end of the pH scale is symptomatic of the acid error. The initial acid error of Corning 015 glass electrodes is 0.17 pH for 3 molar sulfuric acid, 1.7 pH for 12.7 molar sulfuric acid, and 8.9 pH for 18 molar sulfuric acid. Since acid error does not start above 1 pH, acid error is usually not a problem for feedback loops because their setpoint is usually at a much higher pH. However, some pH feedforward loops of strong acid influent have failed due to the acid error. Figure 4-9 shows that the initial pH correction magnitude increases exponentially as the pH decreases below 0.5 pH. The acid error can be reduced by certain glass formulations. For example, a measurement electrode made from Jena glass is error free down to −1.7 pH. For more information on the acid error, the reader is directed to Ref. 17.

A *short* from the measurement to the reference electrode terminals will cause a zero millivolt input and thus a 7 pH reading. In fact, the standby position on some pH meters connects a shorting strap between these terminals to give a 7 pH reading and to prevent polarization of the electrodes during immersion of the electrodes or application of power to the transmitter.

Moisture on the measurement electrode terminal, which decreases resistance R3 in Fig. 4-3, is more of a problem than moisture on the reference electrode terminal, which decreases resistance R4 in Fig. 4-3. Equations (4-8a) and (4-8b) show the error that results from the voltage divider effect of moisture on these terminals. For resistances R3 and R4 reduced to 10 megohms due to moisture, the error for a 10 megohm measurement electrode is 50% while the error for a 10 kilohm reference electrode is only 0.1% (Ref. 18).

$$E^1_m = E_m * \frac{R_2}{R_1 + R_2 + R_3} \qquad (4\text{-}8a)$$

$$E^1_r = E_r * \frac{R_4}{R_4 + R_5 + R_6} \tag{4-8b}$$

where:

E_m = true measurement electrode input voltage (mv)

E^1_m = measured measurement electrode input voltage (mv)

E_r = true reference electrode input voltage (mv)

E^1_r = measured reference electrode input voltage (mv)

R_1 = measurement electrode glass resistance (ohms)

R_2 = measurement electrode internal fill resistance (ohms)

R_3 = measurement electrode terminal resistance to ground (ohms)

R_4 = reference electrode terminal resistance to ground (ohms)

R_5 = reference electrode internal fill resistance (ohms)

R_6 = reference electrode liquid junction resistance (ohms)

4-3. Electrode Response

The time response of a pH measurement electrode without any problems directly immersed in an aqueous solution can be approximated by a time constant whose value varies between two and 30 seconds. The major contributors to this time constant are a boundary layer of fluid around the electrode bulb that slows down the diffusion and migration of ions to the pH sensitive glass and a double layer effect at the surface of the glass that poses an energy barrier to the proton (Ref. 19). The hydronium ion moves from regions of high to low concentration (diffusion) and moves from regions of high to low positive charge (migration). The boundary layer, which slows down this movement, increases in thickness as the velocity of the stream past the electrode decreases. The time constant for velocities below one foot per second is large enough to deteriorate loop performance. Since the velocity in pumped pipelines is between three and seven feet per second whereas the velocity in sample lines and agitated vessels is usually less than one foot per second, the pipeline injector type of electrode described in the next section should be considered whenever a location in a pumped pipeline close to the mixing equipment outlet is available. The double layer has the greatest effect in nonaqueous solutions where the measurement electrode time constant may be several hundred seconds. It is probably responsible for the time constant for a decrease in pH being about twice as large as the time constant for an increase in pH and for the time constant for buffered solutions being much smaller (Ref. 20). The time it takes for the proton to move to its site in the glass once it jumps the double layer energy barrier has been measured to be about 30 milliseconds by using a jet stream to eliminate the boundary layer and double layer effect (Ref. 19). Table 4-4 summarizes some test results on electrodes for different stream velocities, direction, and magnitude of pH changes.

Direction of Change	Magnitude of Change (ΔpH)	Buffering ?	Velocity (fps)	Time Constant (seconds)
Positive	0.5	No	5	1.2
Negative	0.5	No	5	2.8
Positive	1.0	Yes	5	0.25
Negative	1.0	Yes	5	0.5
Positive	1.5	No	5	1.8
Negative	1.5	No	5	6.2
Positive	3.0	Yes	4	0.75
Negative	3.0	Yes	4	1.5
Positive	3.0	Yes	2	1.5
Negative	3.0	Yes	2	3
Positive	3.0	Yes	1	2
Negative	3.0	Yes	1	4
Positive	3.0	Yes	0.5	3
Negative	3.0	Yes	0.5	12

Table 4-4. The electrode time constant depends upon the direction and magnitude of pH change, buffering, and the velocity of the sample.

A change in salt concentration will change the electric field and alter the migration of hydrogen ions enough to change the pH until a concentration gradient develops and the diffusion of the hydrogen ions reestablishes the original hydrogen ion activity at the measurement electrode surface. For example, the addition of 0.1 moles/liter of sodium chloride to a 3 pH solution of hydrochloric acid will cause the pH to temporarily drop to 2.9 pH and return to 3 pH about six seconds later (Ref. 21). The measured pH goes through a transient even though the true pH has not changed.

The time constant and dead time (time delay) due to an electrode flow chamber and the dead time due to a sample line are frequently larger than the electrode time constant. The flow chamber is partially backmixed due to incoming flow turbulence so that the residence time (volume divided by flow) is split between a time constant and an equivalent dead time. If the turbulence creates an agitation flow equivalent to the sample flow, the split is equal so that the time constant and dead time are both equal to one half the residence time per Eq. 4-9. The sample line dead time can be assumed to be equal to the residence time of the sample line due to plug flow per Eq. 4-10 because the small time constant from the velocity profile is usually small compared to the total transportation delay. Since the volume of the sample line is equal to the length multiplied by cross-sectional area, the dead time can be decreased by decreasing the line length or diameter or increasing the sample flow. The same relationship holds for pumped pipeline transportation delay. A common mistake is to install a large agitator to reduce equipment dead time and then install the electrodes in an electrode flow chamber or pot with a long length of intervening sample line or pumped pipeline so that the resulting measurement dead time is greater than the equipment dead time. A heat exchanger in the sample line to lower the sample temperature or in the pipeline to remove the heat of neutralization can add significantly to the flow path length and hence the transportation delay. Units 9 and 10 will discuss how loop performance deteriorates as loop dead time increases.

$$TC_c = TD_c = \frac{V}{2*Q} \tag{4-9}$$

$$TD_l = \frac{A*L}{Q} \tag{4-10}$$

where:
A = cross-sectional area of sample line or pipeline (sq ft)

Q = sample flow rate (cu ft/min)

L = length of sample line or pipeline (ft)

TC_c = time constant of backmixed electrode flow chamber (minutes)

TD_c = dead time (mixing delay) of electrode flow chamber (minutes)

TD_l = dead time (transportation delay) of sample line or pipeline (minutes)

V = volume of electrode flow chamber (cu ft)

4-4. Electrode Installation

The most popular electrode installation assembly until recently was a flow chamber of various sizes and shapes. A small sample stream was diverted from the tank or pipeline to the flow chamber. Strainers, filters, and ultrasonic cleaners were added to reduce coating problems. Frequently the plugging of strainers became as much of a maintenance headache as the coating of the electrodes. The ultrasonic cleaner was only effective for loose particulate coatings. The development of the injector type of electrode assembly shown in Fig. 4-10 allowed the insertion and removal of electrodes into pressurized vessels and pipelines. The high velocities of pumped pipeline flow decreases the coating rate and the time constant of the electrodes. The electrode can be removed even though the pipeline pressure is up to 100 psig by slightly loosening the swagelock fitting that grips the electrode metal sheath, slowly withdrawing the sheath until it clears the ball valve, closing the ball valve, and withdrawing the sheath the remaining distance. In some respects, this type of assembly is safer than a flow chamber with an integral terminal enclosure because the sealing method is easily checked and any leaks are visible. A tee and an additional ball valve can be provided for flushing the assembly before removal of the sheath.

Amico Instruments, Inc., Monitek, Inc., and Van London Company have developed assemblies that reduce coating problems by mechanical wiping of the electrode surface by soft teflon brushes. While this method is effective in removing coatings that will not stick to the teflon brushes, the mechanical assembly requires air or power and maintenance. SensoreX has recently developed an assembly that reduces coating problems by directing the sample stream at high velocity past a large flat surface electrode. This method requires diversion of a sample stream to the assembly. Horiba has developed a jet washing system for a submersion electrode assembly that reduces coating problems by periodically directing a jet of cleaner fluid at the electrode surface. This method requires holding the pH measurement or

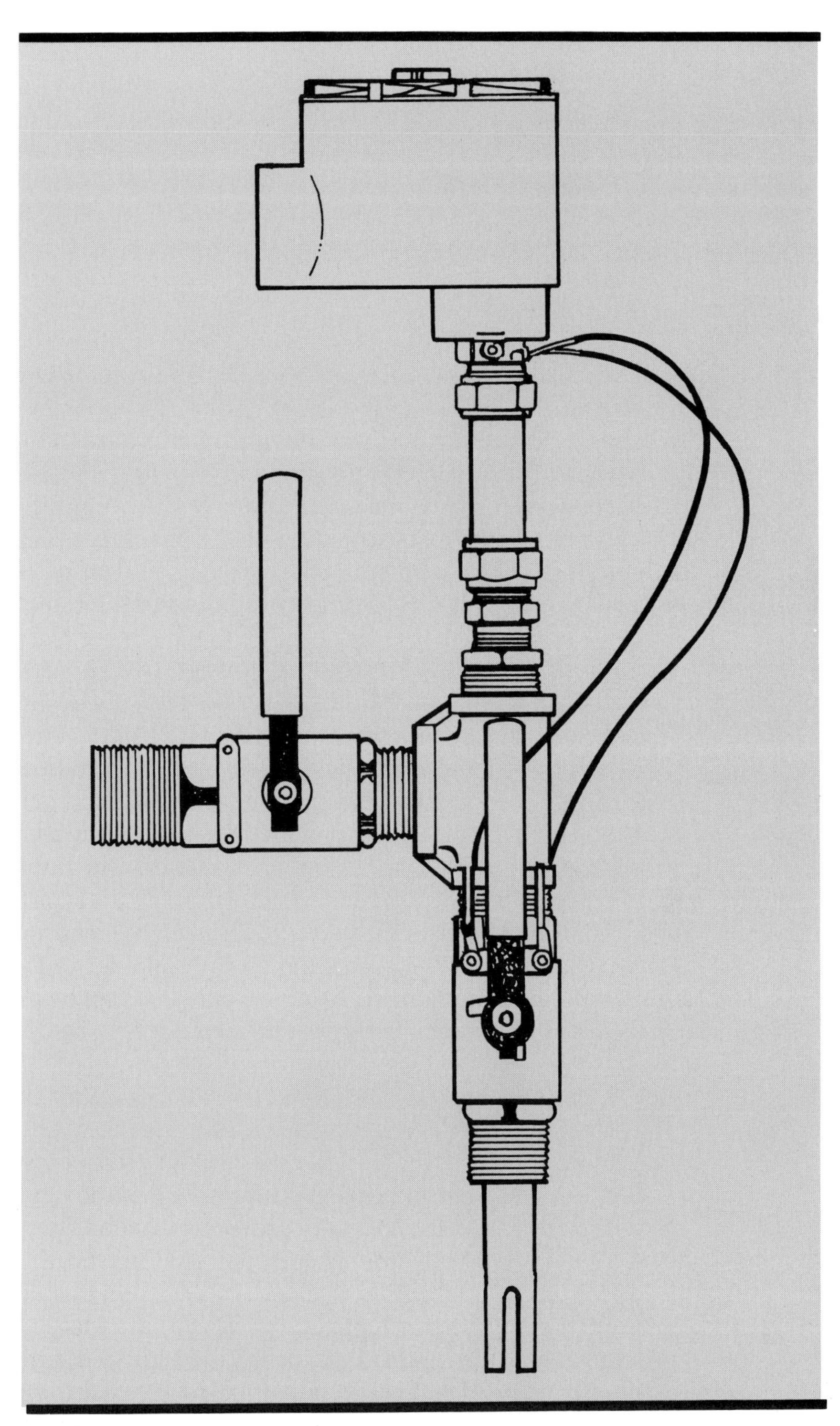

Fig. 4-10. The injector type of electrode assembly reduces the coating rate and time constant of the measurement electrode (Courtesy of the Van London Company, Houston, Texas)

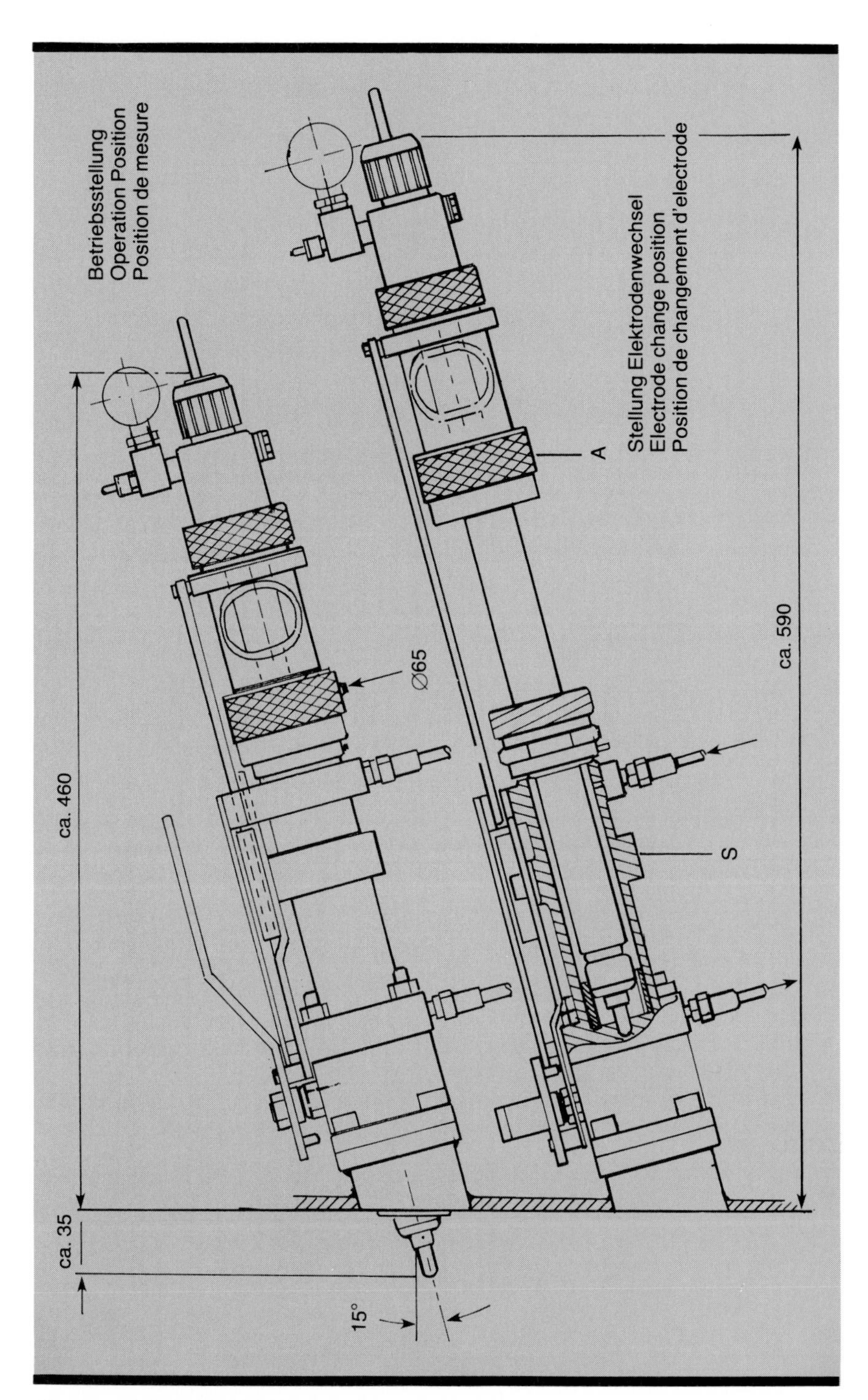

Fig. 4-11. Biochemical and food processes require an electrode assembly that is field sterilizable and has no cavities in which process material can accumulate (Courtesy of Ingold Electrodes, Inc.).

controller output at its last value during the washing and requires appreciable quantities of a cleaner fluid for the washing that is compatible with the process fluid.

Biochemical and food processes require sanitary electrode assemblies that are field sterilizable and have no cavities in which process material can accumulate. The electrodes must have a higher than normal pressure and temperature rating to withstand periodic steam sterilization. Figure 4-11 shows an electrode assembly commonly used for fermentation pH measurement.

The measurement and reference electrode terminal enclosures cannot generally be made tight enough to prevent the moisture accumulation during the night from the decrease in the ambient temperature which creates enough of a vacuum inside the enclosure to pull in moist night air. While desiccant packages may initially alleviate a moisture problem, the absorption of moisture creates a greater vacuum which increases the moisture pulled in until the package is saturated. A wet desiccant package lying against the measurement electrode terminal will provide a lower resistance path to ground than a few drops of water. The best method of preventing moisture accumulation is purging the enclosure with clean dry instrument air (Ref. 22).

Moisture accumulation on electrode terminals in submersion assemblies is a problem despite gasketing and purging. Any terminals or electrode electrical connections below the surface should be sealed in plastic. It is the author's experience that a throw-away, sealed, plastic electrode submersion assembly is more reliable and is cheaper in the long run than an assembly that has submerged terminals to facilitate replacement of just the electrodes.

Exercises

4-1. *If the hydrogen activity is 0.0000001 normality for the internal fill of a measurement electrode without any application problems, at what pH is the isopotential point?*

4-2. *What is the slope of the millivolt versus pH curve if the solution temperature is 100°C and there are no application problems?*

4-3. *What are the potential causes of an erratic pH reading?*

4-4. *What are the potential causes of a slow pH reading?*

4-5. *What is the potential cause of a shortened span only at high end of the pH scale?*

4-6. *What is the potential cause of a constant 5 pH reading?*

4-7. *What are the potential causes of a constant 7 pH reading?*

4-8. *What are the potential causes of a constant 0 pH reading?*

4-9. *What is the sample transportation delay due to installing the electrodes in a two-inch schedule 40 pipeline 50 feet from the effluent discharge nozzle for an effluent flow of 100 gpm?*

4-10. *List the methods employed to reduce electrode coating problems.*

References

[1]Skoog, D. A., and West, D. M., *Principles of Instrumental Analysis*, Saunders College, 1980, pp. 331-333.
[2]Shinskey, F. G., *pH and pION Control in Process and Waste Streams*, John Wiley & Sons, 1973, pp. 18, 19.
[3]Skoog, D. A., and West, D. M., *Principles of Instrumental Analysis*, Saunders College, 1980, pp. 541, 542.
[4]Meriman, D., Van London Co., Inc., July 15, 1983, memo to author.
[5]Shinskey, F. G., *pH and pION Control in Process and Waste Streams*, John Wiley & Sons, 1973, pp. 21.
[6]Wescott, C. C., *pH Measurement*, Academic Press, 1978, pp. 19, 20.
[7]*Ibid.*, pp. 58-60.
[9]*Ibid.*, pp. 59-64.
[9]*Ibid.*, pp. 18, 19.
[10]Bates, R. G., *Determination of pH Theory and Practice*, John Wiley & Sons, 1964, pp. 301-304.
[11]McMillan, G. K., "Guidelines for the Measurement and Estimation of Electrical Conductivity for Chemical Solutions," ISA Conference Paper 705, October 1978, pp. 196-206.
[12]Wescott, C. C., *pH Measurement*, Academic Press, 1978, pp. 128, 129.
[13]Rushton, C. and Bottom, A., "Measuring pH in Low Conductivity Waters," *Kent Tech. Rev.* (G.B.), No. 24, March 1979, pp. 3-5.
[14]Meriman, D., Van London Co., Inc., July 15, 1983, memo to author.
[15]McMillan, G. K., *Tuning and Control Loop Performance*, ISA Monograph 4, 1983, pp. 102-104.
[16]Bates, R. G., *Determination of pH Theory and Practice*, John Wiley & Sons, 1964, pp. 315-319.
[17]*Ibid.*, pp. 319-322.
[18]Shinskey, F. G., *pH and pION Control in Process and Waste Streams*, John Wiley & Sons, 1973, pp. 30, 31.
[19]Hershkovitch, H. Z., et. al., "Dynamic Modeling of pH Electrodes," *The Canadian Journal of Chemical Engineering*, Vol. 56, June 1978, pp. 347-348.
[20]*Ibid.*, pp. 353.
[21]*Ibid.*, pp. 351-352
[22]Shinskey, F. G., *pH and pION Control in Process and Waste Streams*, John Wiley & Sons, 1973, pp. 38-40.

John C. Pfeiffer, P.E.
Pfeiffer Engineering Co.
560 Garden Drive
Louisville, Kentucky 40206

Unit 5:
pH Mixing Equipment

UNIT 5

pH Mixing Equipment

This unit describes the dynamic characteristics of mixing equipment in enough detail so that the effect of equipment type and degree of agitation on pH controller settings and loop performance can be estimated in Unit 9 and the proper equipment selected in Unit 10.

Learning Objectives—When you have completed this unit you should:

A. Know how to estimate the time constant and dead time for any type of mixing equipment.

B. Recognize the limitations on the quality of agitation imposed by the type of mixing equipment.

5-1. Mixing Dynamics

The objective for pH control is to disperse a localized concentration of an acid or base throughout the whole volume as quickly as possible. It is important to remember that extremely small changes in acid or base concentration correspond to large pH changes for most pH setpoints. There are many different internal flow patterns from agitation and many different parameters to quantify the amount of agitation. For pH control, the flow pattern should be axial, as shown in Fig. 5-1a in which the fluid is pulled from the top on down near the shaft and circulated out along the bottom to the sidewalls and back up to the top near the sidewalls. The pattern is called axial because of the vertical up-and-down flow pattern parallel to the axis of the shaft. Baffles that are 90 degrees apart extend vertically along the entire length of the sidewall and are one-twelfth the diameter in width are recommended because they help establish the vertical flow currents, increase the uniformity of the flow pattern, reduce vortexing, reduce swirling, reduce air induction from the surface, and assure a more consistant power draw. Tests made with a pitched blade turbine showed that the addition of baffles reduced the time for complete dispersion (mixing time) from 28 to 20 seconds and reduced the standard deviation from 4.8 to 2.4 seconds (Ref. 1). Propeller and pitched blade turbines provide an axial flow pattern. A double spiral blade and a tangential jet nozzle cause a corkscrew axial pattern that is undesirable because the concen-

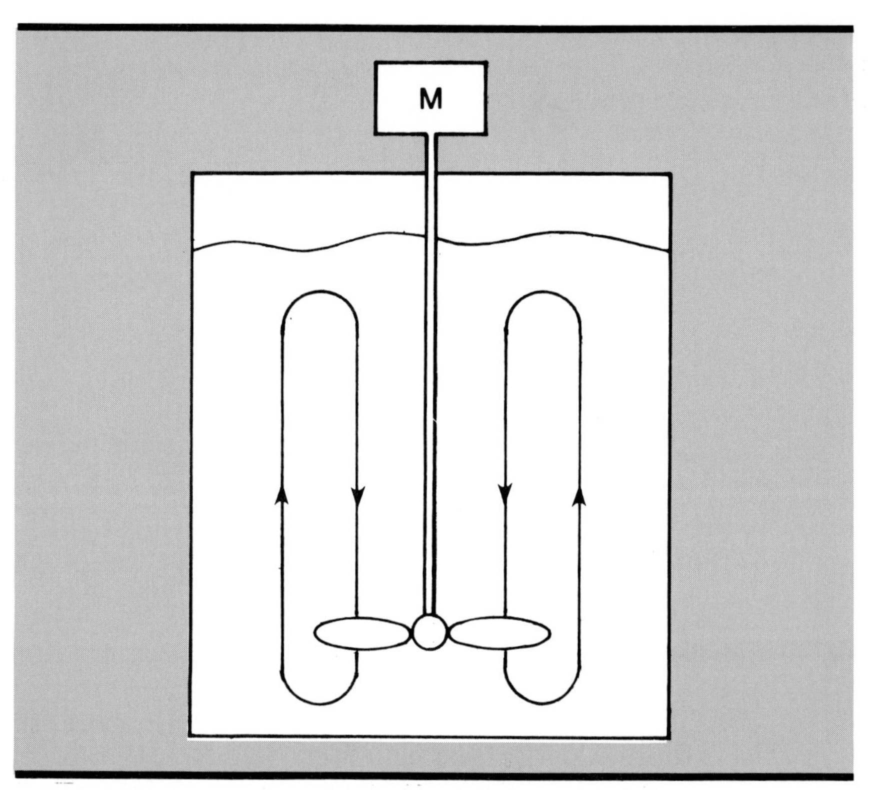

Fig. 5-1a. The axial flow pattern provides the quickest and most complete mixing required for pH control.

tration change has a long and slow corkscrew flow path to follow before complete dispersion. A flat blade or bar turbine cause a radial flow pattern (as shown in Fig. 5-lb) in which the fluid is pushed and pulled radically out from the axis of the agitator shaft. The radial pattern is undesirable because it results in an area of fluid near the surface that is not mixed as vigorously as required for pH control. A side entry axial agitator will produce the undesirable radial flow pattern at the point of entry. Side entry agitators should be avoided.

The equipment dead time, which is the time to the start of the change in equipment outlet concentration after a change in the inlet concentration, is approximately equal to the volume divided by the summation of the total input flow and the agitator pumping rate per Eq. (5-1a). The total input flow is the summation of the influent and reagent flows. For continuous equipment, the effluent flow equals the summation of the influent and reagent flows by level control or overflow. For batch operation, the effluent flow is zero except at the completion of the batch.

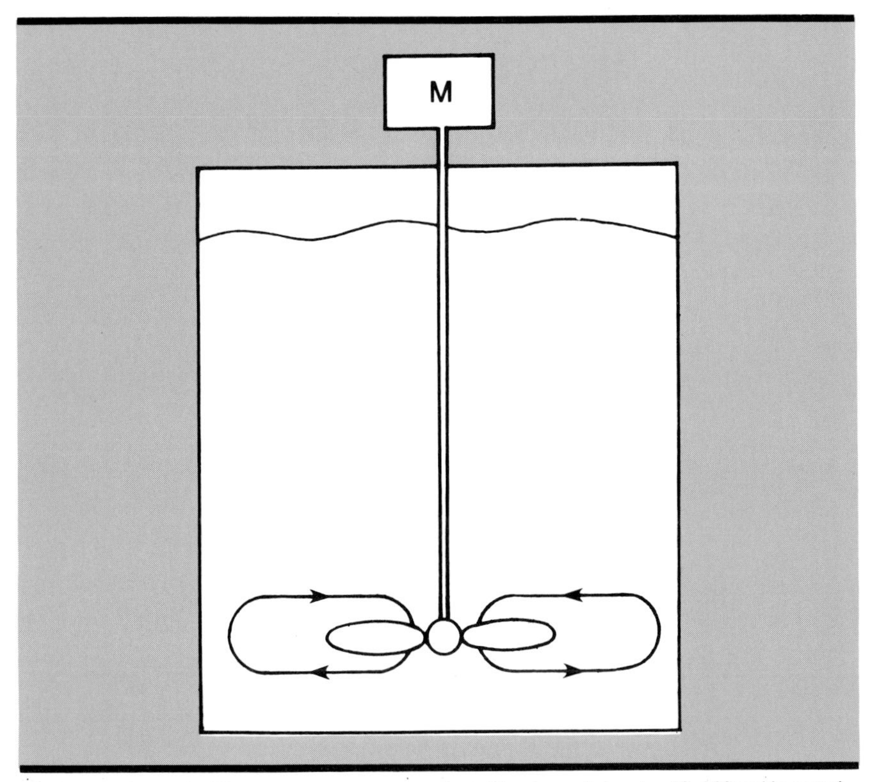

Fig. 5-1b. The radial flow pattern causes a region of inadequately mixed fluid for pH control near the top.

Equation (5-1a) is valid for batch operation as well as continuous operation except the batch volume increases with time per Eq. (5-1d). The agitator pumping rate can be estimated as the product of the discharge coefficient, speed, and the third power of the impeller diameter per Eq. (5-1b). The discharge coefficient can be approximated by Eq. (5-1c) and normally ranges from 0.4 to about 0.8. The impeller diameter in these equations is the diameter of the circle traced out by the tips of the propeller or turbine blades. These equations assume axial flow throughout the entire contents of the vessel. The lower limit on equipment dead time depends on what degree of agitation starts air entrainment from the surface and what the distance is from the inlet to exit nozzles. The dead time cannot be less than the shortest distance between the inlet and exit divided by the internal circulation velocity per Eq. (5-1e). The internal circulation velocity rarely exceeds one foot per second. The largest of the equipment dead time values calculated from Eqs. (5-1a) and (5-1e) should be used as the equipment dead time. Note that this dead time does not include any transportation delay from the equipment exit to the pH electrodes. If only the "turnover time" is specified for an agitator, it can be used as a conservative estimate of the equipment dead time.

$$TD_p = \frac{V}{Q_a + Q_t} \qquad (5\text{-}1a)$$

$$Q_a = N_q * N * D^3_i \qquad (5\text{-}1b)$$

$$N_q = \frac{0.4}{\left[\frac{D_i}{D_t}\right]^{0.55}} \qquad (5\text{-}1c)$$

$$V = Q_t * T + V_O \quad \text{(for batch operation)} \qquad (5\text{-}1d)$$

$$TD_p > \frac{L}{U} \qquad (5\text{-}1e)$$

where:
D_i = impeller diameter (feet)

D_t = tank inside diameter (feet)

Q_a = agitator pumping rate (gpm)

Q_t = total input flow rate (gpm)

L = distance from entrance to exit nozzle (feet)

N = discharge coefficient of impeller (0.4 to 0.8)

Nq = impeller speed (rpm)

T = time since start of the batch (minutes)

TD_p = equipment dead time (minutes)

U = internal circulation velocity (fps)

V = vessel working volume (gallons)

V_o = vessel volume at the start of the batch (gallons)

(a "p" subscript is used instead of a "m" subscript to designate a mixing process parameter because a "m" subscript will be used in Units 9 and 10 to designate a measurement parameter)

The equipment time constant, which is the time it takes the outlet concentration to reach 63% of its final value after the equipment dead time for a step change in the inlet concentration, can be approximated as the residence time minus the equipment dead time per Eq. (5-2). This equipment time constant is a negative feedback time constant for the continuous mode of operation where there is a continuous effluent flow. It is important to realize that this definition is for a linear change in outlet concentration. A pH reading cannot be used to measure this time constant because of its nonlinear relationship to acid or base concentration.

The agitation term "mixing time" refers to the time required for the entire contents of the vessel to reach within a specified percent of the final concentration for a pulse of reagent or influent and no effluent flow (batch mode of operation). This term cannot be related to and should not be confused with the negative feedback time constant for the continuous mode of operation. The "mixing time" can be used to estimate the equipment time constant for the batch mode if the percent uniformity reached is defined. Equation (5-3a) shows that the batch equipment time constant is approximately equal to one-fifth of the mixing time to reach 99% uniformity minus the equipment dead time. The same agitator data sheet should be used to get the "mixing time" and "turnover time" and the "turnover time" should be used as the equipment dead time. It takes about 10 to 15 "turnover times" to achieve a 99% uniformity for an aqueous solution and an axial flow turbine agitator. This translates to a dead time to time constant ratio of 0.3 to 0.5 for a well-mixed batch vessel. Of course, a negative value for the equipment time constant is not physically possible and is a flag of data inconsistency.

The agitation term "circulation time" is the time interval between consecutive passes of a particle through the impeller. The "circulation time" is less than the "mixing time" and is not a good estimate of the degree of mixing because a particle could be recycled back through the impeller region without mixing with the other regions.

For batch equipment, the residence time and hence the negative feedback equipment time constant is infinite since the concentration never reaches a final value for a step change but ramps until a physical limit such as tank overflow or a interlock limit such as high-level shutdown is reached. The slope of the ramp is the integrator gain. This gain depends upon what is manipulated variable, what is controlled variable, and the units of both variables. Equation (5-3b) gives the integrator gain for a controlled variable that is the reagent acid or base weight fraction and a manipulated variable that is reagent mass flow rate. The actual controlled variable is pH but the effect of the nonlinear gain will be included in Units 9 and 10 by translating a change in the abscissa to a change in the ordinate of the titration curve. For continuous operation, the abscissa is the ratio of reagent mass flow to influent mass flow. For batch operation, the abscissa simplifies to reagent weight fraction if the influent mass flow addition was completed before the start of the reagent addition and the reagent mass to be added is small relative to the total mass.

For continuous mode of operation:

$$TC_p = \frac{V}{Q_t} - TD_p \qquad (5\text{-}2)$$

For batch mode of operations:

$$TC_p = \frac{T_m}{5} - TD_p \qquad (5\text{-}3a)$$

$$K_i = \frac{X_{rr} * F_r}{M_b} \qquad (5\text{-}3b)$$

where

K_i = integrator gain for batch pH control (1/min)

M_b = mass of fluid in batch (1b)

Q_t = total input flow (gpm)

T_m = mixing time to reach 99% of final concentration (minutes)

TC_p = equipment time constant (minutes)

TD_p = equipment dead time (minutes)

V = vessel working volume (gallons)

X_{rr} = weight fraction of reagent in the reagent stream

Even though the dynamic response of an agitated vessel is approximated as the combination of an equipment dead time and time constant as shown in Fig. 5-2a, the actual response is S-shaped, as shown in Fig. 5-2b for a step change in concentration. No real process has the sharp transition shown at the end of the dead time in Fig. 5-2. The equipment dead time constant can be graphically approximated by constructing a tangent as shown in Fig. 5-2b. The time from the start of the step change to the intersection of the tangent with the initial value is the dead time and the time from the end of the dead time to the intersection of the tangent with the final value is the time constant. If a field recording of pH is used, the resulting dead time and time constant includes the effect of instrument dynamics and titration curve nonlinearity.

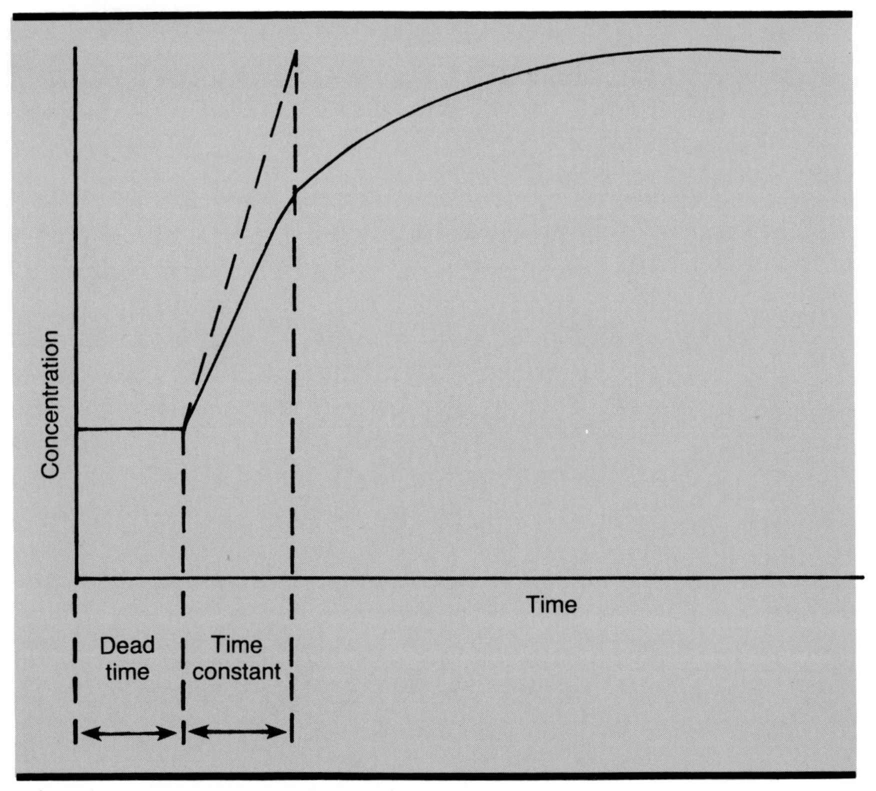

Fig. 5-2a. The theoretical response due to an equipment dead time and time constant for a step input change has a sharp transition at the end of the dead time.

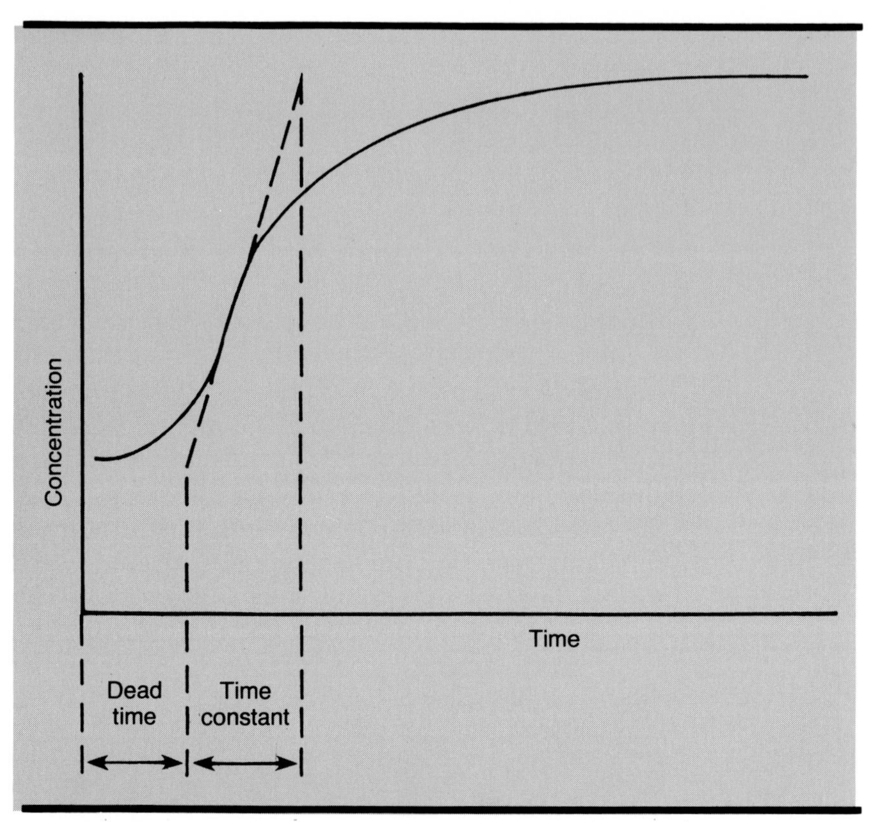

Fig. 5-2b. The actual response due to an equipment dead time and time constant for a step input change has a smooth S-shaped curve.

5-2. Agitated Vessels

The entire contents of a vessel can be axial mixed only if the height of the vessel is equal to or slightly greater than the width. For a single impeller, the height should be less than 150% of the width. This type of vessel will hereafter be referred to as a vertical tank. Equation (5-1) for the equipment dead time is only valid for a vertical tank. If the ratio of the equipment dead time to the time constant is equal to or less than 0.05, the vessel is classified as a vertical well-mixed tank. The importance of this dead time to time constant ratio for controller tuning and control loop performance will be discussed in Units 9 and 10. Horizontal tanks have a length much greater than the height. No matter how many agitators are installed, the complete volume cannot be considered as axially mixed. There will be regions of stagnation, short circuiting, and plug flow. The result of stagnation is a region of fluid whose concentration is only dispersed by ion diffusion or migration. The result of short circuiting is a region where a change in reagent or influent concentration bypasses the rest of the vessel contents and appears in the effluent prematurely. The result of plug flow is a region where the equipment time constant approaches zero and the dead time to time constant ratio approaches infinity. Figure 5-2 illustrates a stagnant region in the far right corner of the horizontal tank due to no agitation or entry flow, a short-circuiting region in the far left corner due the proximity of the reagent and the effluent nozzles, and a plug flow region of influent to effluent in the middle of the tank due to inadequate axial agitation.

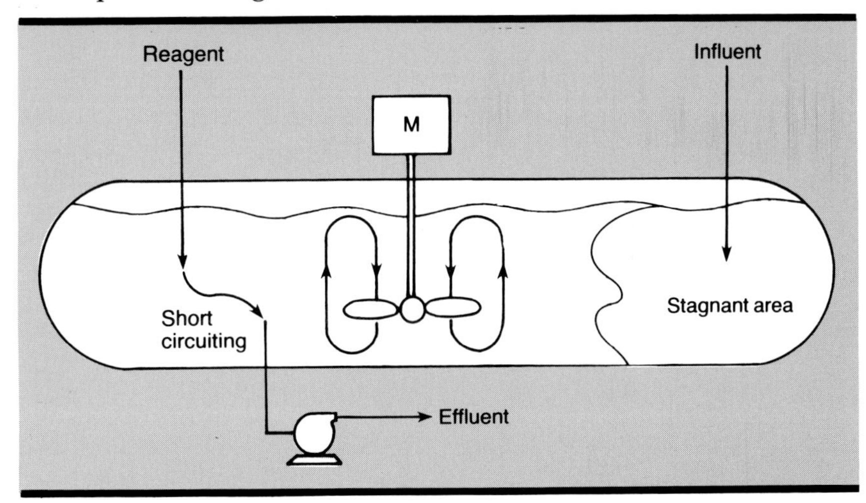

Fig. 5-3. A horizontal tank may have regions of stagnation, short circuiting, and plug flow.

The equipment dead time in a horizontal tank with a mechanical agitator will typically vary between 20 to 40% of the residence time as shown by Eq. (5-4) if there is no short circuiting. If there is no mechanical agitation, the equipment dead time varies from 20 to 30% of the residence time for a vertical tank due to the dispersion from flow entry and exit turbulence, ion diffusion, and ion migration. The dead time percent of residence time for an unagitated horizontal tank proportionally increases with the ratio of length to height (Ref. 2). Such data are important when considering the effectiveness of an unagitated vessel as a filter of disturbances or control loop oscillations since the filter time constant is equal to the equipment time constant, which is the portion of the residence time not converted to dead time per Eq. (5-2). Figure 5-4 shows that the attenuation of an oscillation amplitude is greater if a given volume is split into two volumes in series if the equipment time constant is larger than the oscillation period.

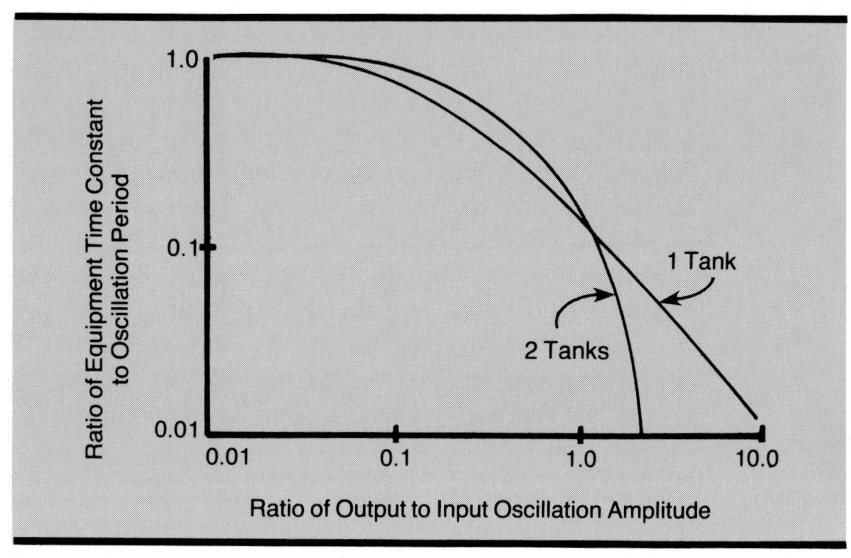

Fig. 5-4. A given volume can be more effectively used to filter oscillations by splitting it up into two individual volumes in series if the equipment time constant of each individual volume is greater than the oscillation period.

For a horizontal tank with a mechanical agitator:

$$0.20 * \frac{V}{Q_t} < TD_p < 0.40 * \frac{V}{Q_t} \tag{5-4}$$

where:

Q_t = total input flow (gpm)

TD_p = equipment dead time (minutes)

V = vessel working volume (gallons)

5-3. Static Mixers

Static mixers have motionless internal elements, as shown in Fig. 5-5, that subdivide and recombine the flow stream repeatedly and cause rotational circulation of the flow stream to provide radial mixing of the stream, but very little axial or back mixing, as shown in Fig. 5-6. The equipment dead time is about 75% of the residence time for a static mixer, as shown by Eq. (5-5) (Ref. 3). Flow pulses due to a positive displacement reagent pump and drops due to a high viscosity reagent or low reagent velocity will not be back mixed and will cause a noisy pH signal. Bubbles from a gaseous reagent will also cause a noisy pH signal because the residence time is not sufficient for complete reagent dissolution. The use of additional motionless elements beyond the normal number does not usually increase drop or bubble dispersion as much as an increase in stream velocity by a decrease in static mixer diameter. One advantage of the small residence time is that the magnitude of the equipment dead time is small.

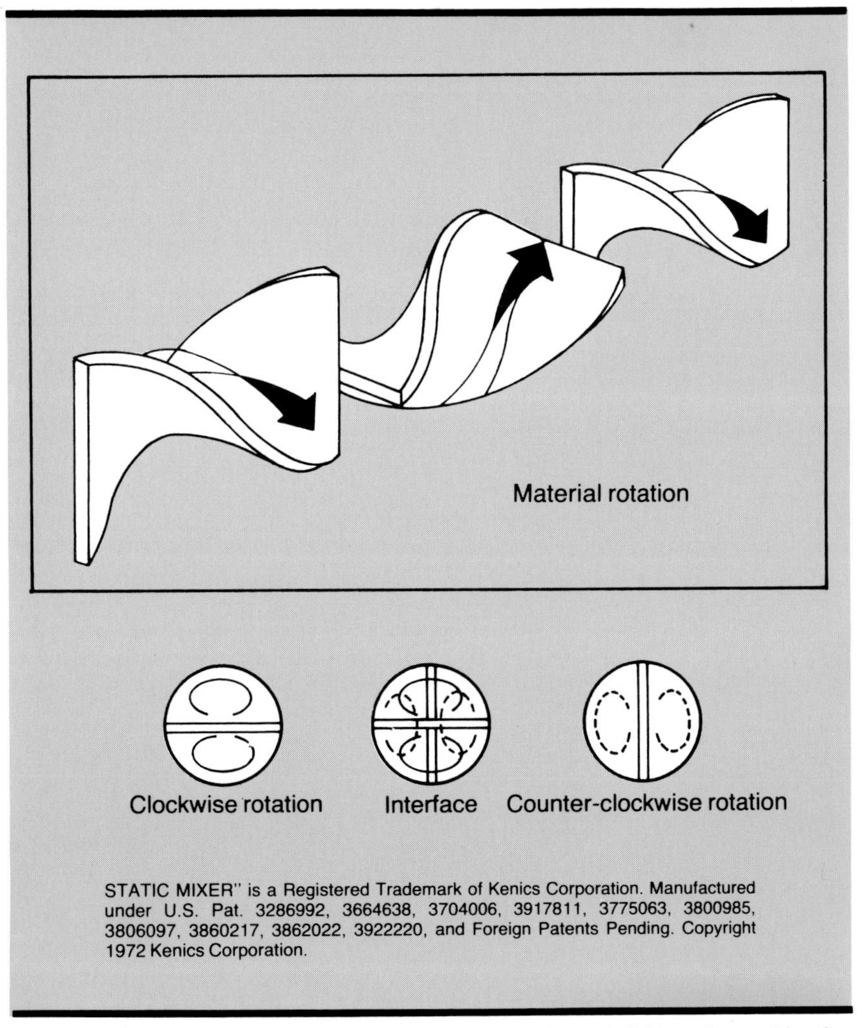

Fig. 5-5. A static mixer has motionless internal elements that subdivide and recombine the flow stream repeatedly and cause rotational circulation.

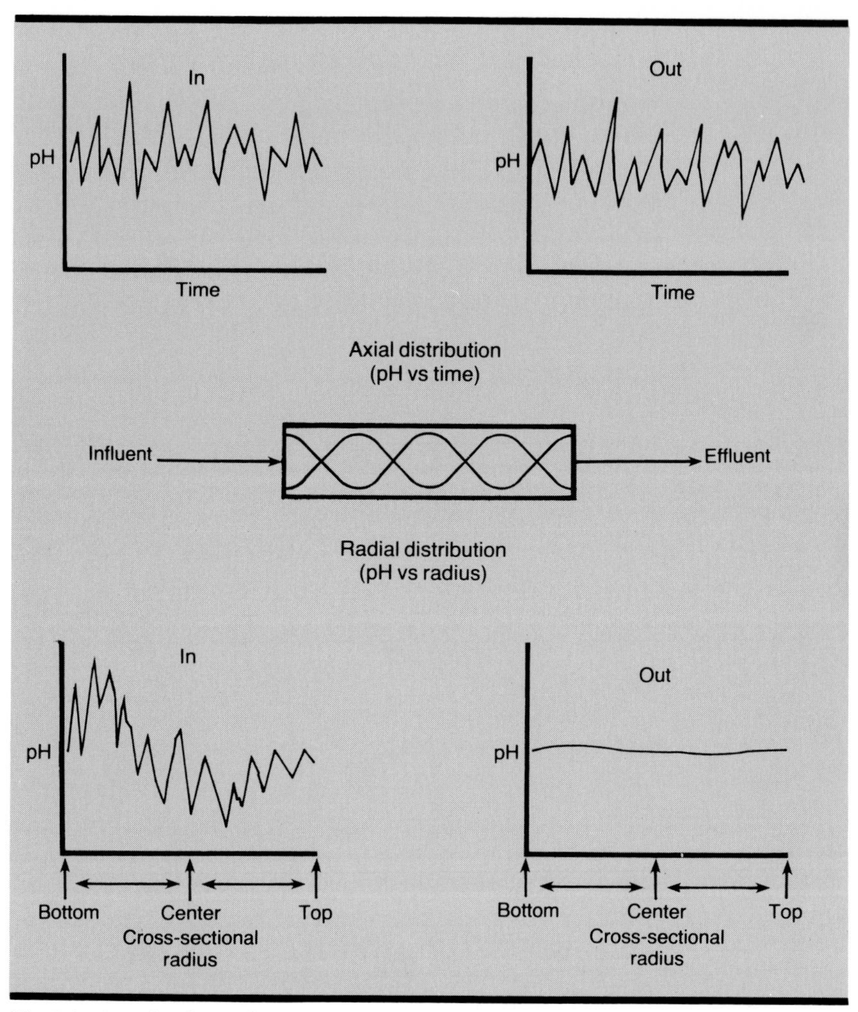

Fig. 5-6. A static mixer will cause a noisy pH signal because there is very little axial or back mixing.

For a static mixer:

$$TD_p = 0.75 * \frac{V}{Q_t} \tag{5-5}$$

where:

Q_t = total input flow (gpm)

TD_p = equipment dead time (minutes)

V = static mixer volume (gallons)

About 100 equivalent pipe diameters will provide about the same degree of radial mixing as a static mixer. However, the equipment dead time is approximately equal to 100% of the residence time as shown by Eq. (5-6) because the equipment time constant from the velocity profile is negligible compared to the transportation delay. The length of piping between the reagent injection point and pH measurement point, and hence the transportation delay, can be reduced by the use of fittings and valves that have a large number of equivalent pipe diameters.

For a pipeline:

$$TD_p = \frac{V}{Q_t} \tag{5-6}$$

where:

Q_t = total input flow (gpm)

TD_p = equipment dead time (minutes)

V = pipeline volume (gallons)

5-4. Sumps, Ponds, and Lagoons

As the width or length of an agitated sump increases with respect to its height, the quality of agitation moves from that for a horizontal tank to that for a pond or lagoon. The equipment dead time will increase from 40 to 80% of the residence time, as shown by Eq. (5-7). A pond or lagoon will have an equipment dead time that increases with size from 80 to 99% of the residence time, as shown by Eq. (5-8) (Ref. 4). There will be stagnation areas in which sizes change with the amount of sunshine and wind. Submerged static mixers with air jets and submerged turbine agitators can provide localized backmixing (Ref. 5). While the portion of the residence time that is dead time is decreased, the magnitude of the residence time, and hence the dead time, is extremely large. The detrimental effects of such a large dead time on controller tuning and control error will be detailed in Units 9 and 10.

For an agitated sump:

$$0.40 * \frac{V}{Q_t} < TD_p < 0.80 * \frac{V}{Q_t} \tag{5-7}$$

For a pond or lagoon:

$$0.80 * \frac{V}{Q_t} < TD_p < 0.99 * \frac{V}{Q_t} \tag{5-8}$$

where:

Q_t = total input flow (gpm)

TD_p = equipment dead time (minutes)

V = sump, pond, or lagoon working volume (gallons)

Exercises

5-1. *What is the equipment dead time and time constant for a vertical tank if the impeller diameter is two feet, the inside tank diameter is four feet, the agitator speed is 400 rpm, the working volume is 1000 gallons, and the total input flow is 100 gpm?*

5-2. *For a given unagitated volume, which type of equipment provides the most effective filtering action for upstream disturbances?*

5-3. *Which type of equipment has the smallest dead time?*

5-4. *Which type of equipment has the largest dead time?*

References

[1]Tatterson, G. B., "Effect of Draft Tubes on Circulation and Mixing Times," *Chem. Eng. Commun.*, Vol. 19, 1982, pp. 146, 147.
[2]Shinskey, F. G., *pH and pION Control in Process and Waste Streams*, John Wiley and Sons, Inc., 1973, p. 161.
[3]Bor, T., "The Static Mixer as a Chemical Reactor," *British Chemical Engineering*, Vol. 16, No. 7, 1971, p. 610.
[4]Moore, R. L., *Neutralization of Waste Water by pH Control*, ISA Monograph 1, 1978, p. 122.
[5]*Ibid.*, p. 125.

Unit 6: pH Control Valves

UNIT 6

pH Control Valves

This unit shows how to estimate the control valve hysteresis, rangeability, and speed requirements to meet a control objective and how to evaluate the alternatives to meet those requirements.

Learning Objectives—When you have completed this unit you should:

A. **Understand how control valve precision and rangeability depend on control valve hysteresis.**

B. **Recognize when control valve stroking time is important.**

C. **Be aware of the advantages and disadvantages of various types of control valves for pH control.**

6-1. Precision

The maximum desired positive and negative excursions of pH about setpoint form a control band. The width of this control band should be marked on the ordinate of the titration curve and translated to a width on the abscissa that is then the allowable reagent error band per unit of influent flow. This error should then be multiplied by the maximum influent flow, divided by the reagent valve capacity, and multiplied by 100% to express the error in terms of a percent of reagent control valve capacity. The reagent valve capacity should be about 25% greater than the abscissa at the center of the reagent error multiplied by the maximum influent flow to allow for data error. Figures 6-1a and 6-1b show how a steeper slope at setpoint requires the reagent error to be smaller, and hence the control valve to be more precise. If Eq. (6-1b) for the maximum reagent valve capacity is substituted into Eq. (6-1a) for the allowable reagent error in percent of capacity, the result is Eq. (6-1c) that has only the graph parameters "A" and "B." Equation (6-1c) shows the reagent error in percent of capacity decreases for an increase in the distance "A" of the center of the reagent error band on the abscissa from the origin. Thus, a lower influent pH requires a smaller allowable reagent error in percent of capacity because the origin, which is the point of zero reagent flow, is shifted to the left, further away from the reagent error band.

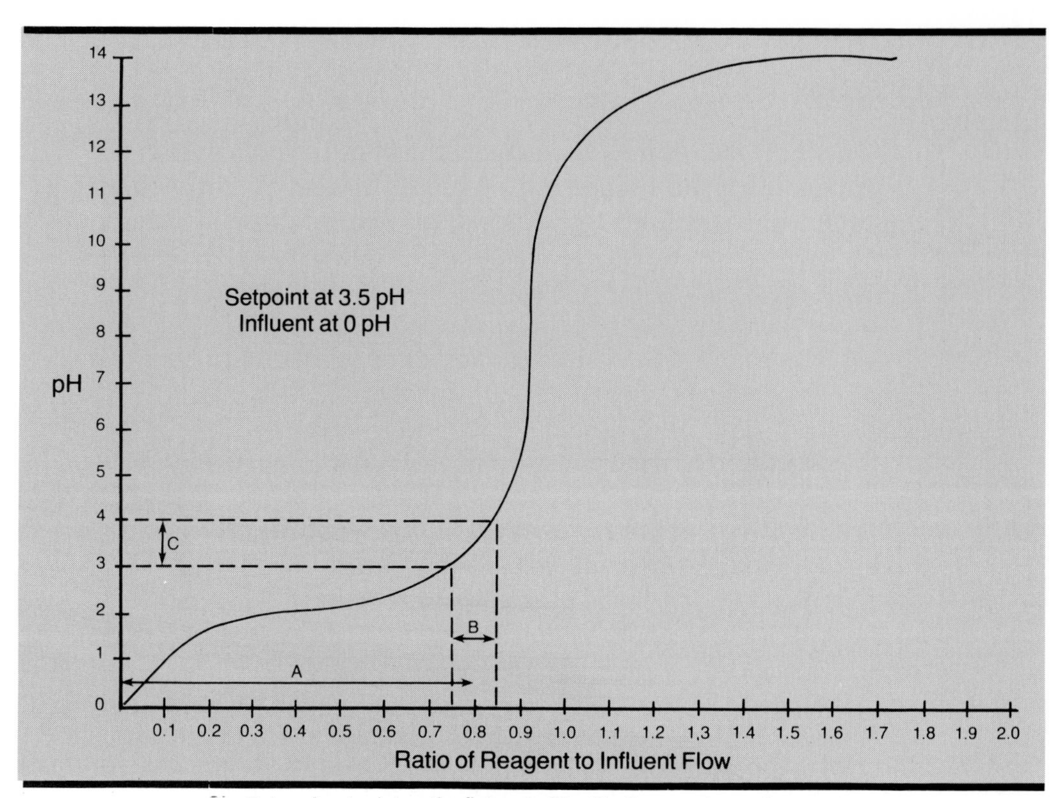

Fig. 6-1a. A setpoint on the flat part of the titration curve translates a given control band to a larger allowable reagent error band.

$$E_r = 100 * F_{imax} * \frac{B}{F_{rmax}} \tag{6-1a}$$

$$F_{rmax} = 1.25 * A * F_{imax} \tag{6-1b}$$

$$E_r = 80 * \frac{B}{A} \tag{6-1c}$$

where:

A = distance of center of reagent error band on abscissa from origin

B = reagent error band per influent flow

E_r = allowable reagent error in percent of reagent valve capacity (%)

F_{imax} = maximum influent flow (pph)

F_{rmax} = reagent control valve capacity (pph)

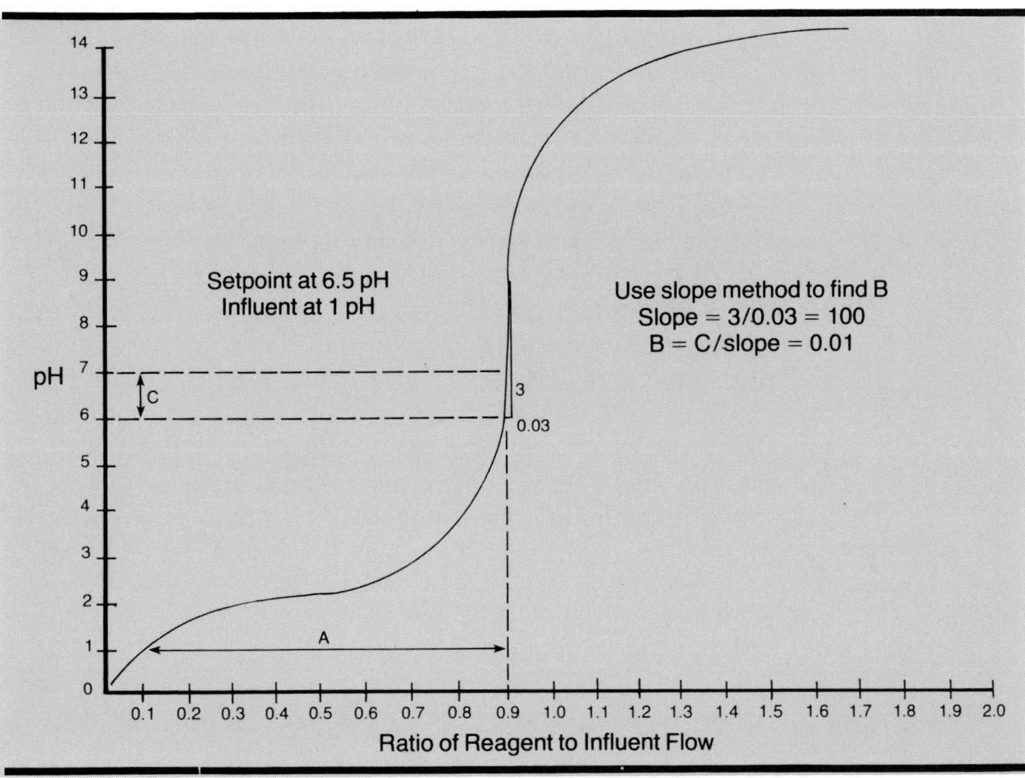

Fig. 6-1b. A setpoint on the steep part of the titration curve translates a given control band to a smaller allowable reagent error band.

The reagent control valve hysteresis must be less than the allowable reagent error as a percent of capacity, as shown by Eq. (6-2) in order for the pH to stay within the control band, even if there are no disturbances. Hysteresis will result in dither about the reagent error band, and hence the control band, if the pH controller has reset action. For a strong acid and base, a 1% hysteresis can cause a ± 3 pH error or a 6 pH control band. Control valve hysteresis is defined by SAMA standard PMC 20.1-1973 as dead band plus hysteretic error. Hysteretic error is the difference in stroke for an increasing versus decreasing signal. It is due to energy absorption in the form of heat by the actuator spring and instrument air. Dead band error is the change in signal required to start the stroke from a stationary position or to change the stroke direction upon a change in signal direction. It is due to friction in the bushings and packing. Normally, the dead band error is larger than the hysteretic error. Figures 6-2a, 6-2b, and 6-2c show the hysteretic error, dead band error, and hysteresis, respectively. Hysteresis values quoted by the manufacturer are for a packing that is only "hand tight." Since most reagents are corrosive and hazardous, the packing of the field valve will have to be much tighter and the hysteresis values greater. Table 6-1 gives typical hysteresis values for "hand tight" packing on a sliding stem valve. Note that the hysteresis is very large for high-friction packings such as laminated graphite (grafoil) and that a 1% hysteresis is only achievable by the use of a positioner. Rotary valves generally have larger hysteresis values (the camflex valve is the most notable exception). Eccentric disk valves also have an overshoot problem when opening from the closed position due to the torque requirement plummeting from its high breakaway value to a low value when it breaks free (Ref. 1). The rotary valve capacity is usually much greater than the maximum reagent flow requirement. If an oversized rotary valve is used, the actual reagent error band is greater than desired since this error band is in terms of percent of valve capacity. Rotary valves should generally be avoided for throttling reagents for pH control.

$$E_v < E_r \tag{6-2}$$

where:

E_r = allowable reagent error in percent of reagent valve capacity (%)

E_v = reagent control valve hysteresis (%)

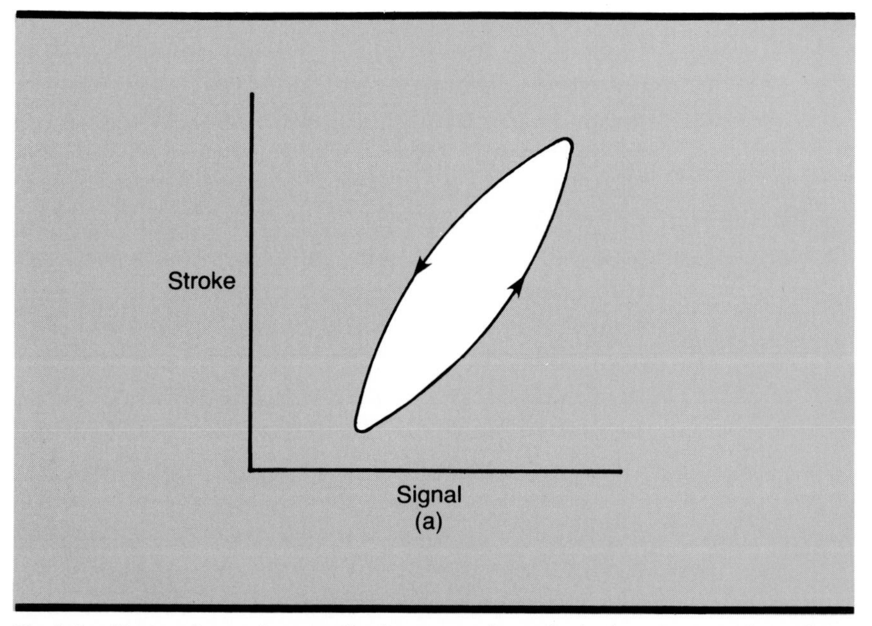

Fig. 6-2a. Hysteretic error is caused by the energy absorption by the actuator spring and air pressure.

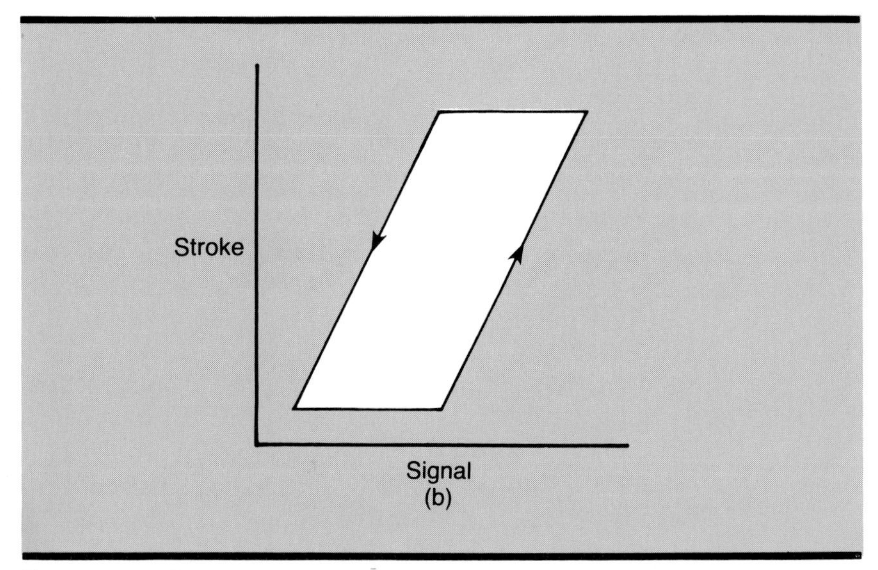

Fig. 6-2b. Deadband error is caused by bushing and packing friction on stem movement.

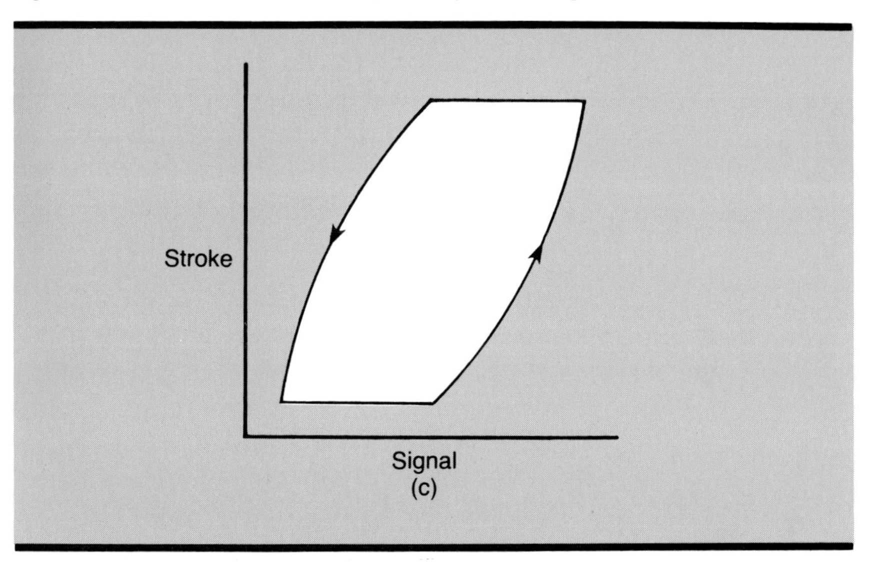

Fig. 6-2c. Hysteresis is the combination of a hysteretic and dead band error.

Packing Type	Positioner	Signal (PSIG)	Hysteresis (%)
TFE	No	3-15	5
TFE	No	6-30	5
TFE asbestos	No	3-15	10
TFE asbestos	No	6-30	7
Laminated graphite	No	3-15	15
Laminated graphite	No	6-30	10
Any of above	Yes	Any of above	1

Table 6-1. Control valve hysteresis for sliding stem valves depends on packing type, pressure signal range, and whether a positioner is used.

6-2. Rangeability

Control valve inherent rangeability is the ratio of the maximum to the minimum controllable flow coefficient within which the inherent flow characteristic does not exceed some specified limits. The inherent flow characteristic is the relationship between flow and travel as the travel is varied between 0 and 100% for a fixed inlet pressure and pressure drop. These definitions are as stated in Ref. 2, except for the addition of the words "controllable" and "inlet pressure." It is advisable to include the word "inlet pressure" because gas flow also depends upon the inlet pressure. It is advisable to include the word "controllable" because the allowable limits on the deviation of the flow characteristic should be based on the magnitude of the change in slope, and hence the valve gain seen by the controller, and because valve stroke precision near the closed position places a practical limit on the minimum controllable flow coefficient. While the flow coefficient of some valves becomes very irregular below 5% opening, the change in slope is still usually less than the change in slope over the whole stroke for an equal percentage trim, a modified parabolic trim, or a varying pressure drop. Up to now, the emphasis has been on trying to define specified limits on the inherent flow characteristic and the effect of stroke hysteresis and overshoot on rangeability has been ignored in the literature. If the effect of stroke precision is ignored, the quick-opening inherent flow characteristic yields the best, the linear inherent flow characteristic the next best, and the equal percentage inherent flow characteristic the worst inherent rangeability. However, if the effect of stroke precision is included, and the valve stroke precision is worse than 2%, the opposite is true. Table 6-2 shows that the equal percentage inherent flow characteristic gives a 50:1 inherent rangeability for a wide range of valve stroke precision values due to the flatness of the characteristic curve near the closed position. Most field instrument maintenance personnel have felt that equal percentage trim gave them the greatest rangeability despite what the literature said. The rangeability required depends upon the maximum and minimum influent pH and flow. The maximum influent pH corresponds to a minimum "A" distance and the minimum influent pH corresponds to a maximum "A" distance. The rangeability required is then the ratio of maximum to minimum "A" distance multiplied by the ratio of maximum to minimum influent flow as shown in Eq. (6-3a) and multiplied by 1.25 to provide a 25% margin for data error. The minimum "A" used should not be less than the reagent error band since the pH is already within the control band. Equation (6-3b) shows that the rangeability for this case where the minimum "A" is equal to the "B" parameter is equal to the inverse of the allowable reagent error as a fraction (not percent) of the reagent valve capacity per Eq. (6-1a) multiplied by the influent flow ratio. Thus, if you have satisfied the hysteresis requirement, you normally have satisfied the rangeability requirement since the actual "A" ratio is normally much less than that assumed for Eq. (6-3b). The literature to date has focused on the rangeability requirement, which is the need to deliver a small change in reagent flow only at low loads, and ignored the more stringent precision requirement, which is the need to deliver a small change in reagent flow at all loads. Therefore, split-ranged valves solve only the rangeability problem whereas high-performance positioners or parallel valves solve both precision and rangeability problems.

$$R_r = 1.25 * \frac{A_{max}}{A_{min}} * \frac{F_{imax}}{F_{imin}} \qquad (6\text{-}3a)$$

for $A_{max} = A$ and $A_{min} = B$:

$$R_r = 1.25 * \frac{A}{B} * \frac{F_{imax}}{F_{imin}} \qquad (6\text{-}3b)$$

where:

A_{min} = minimum distance from influent pH to center of reagent error band

A_{max} = maximum distance from influent pH to center of reagent band

F_{imin} = minimum influent flow (pph)

F_{imax} = maximum influent flow (pph)

While the split ranging of control valves does not solve the precision problem, it is still required for extreme variations in the influent flow or where the influent pH can be above or below the setpoint so that both an acid and base reagent are needed. The split ranging must be done accurately to prevent overlapping of strokes which alters the combined valve gain seen by the controller and which wastes reagent when an acid and base reagent valve are both open. To avoid wasted reagent, a split-range gap of a few percent is used to insure that the acid and base valves are never both open even if a calibration error develops or if a timing problem develops due to differences in stroking speed for different valves or different stroke directions. While the split ranging of the valves can be accomplished by calibrating the I/P transducers or positioners for the split signals, the accuracy of these mechanical devices tends to deteriorate more, and the convenience of accuracy verification is less than that for control room analog modules or functional blocks in microprocessor controllers. Figure 6-3a shows the pH drift that develops from too large a gap of split-ranged acid and base valves and Fig. 6-3b shows the sawtooth oscillation that develops from an overlap of split-ranged acid and base valves.

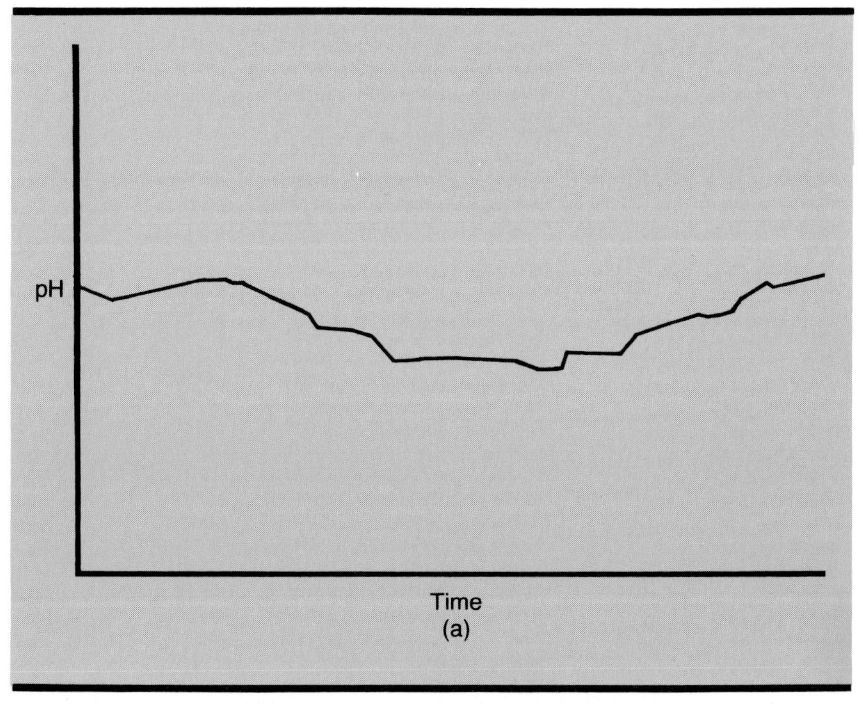

Fig. 6-3a. Too large of a gap in the split range of an acid and base reagent valve causes pH drift.

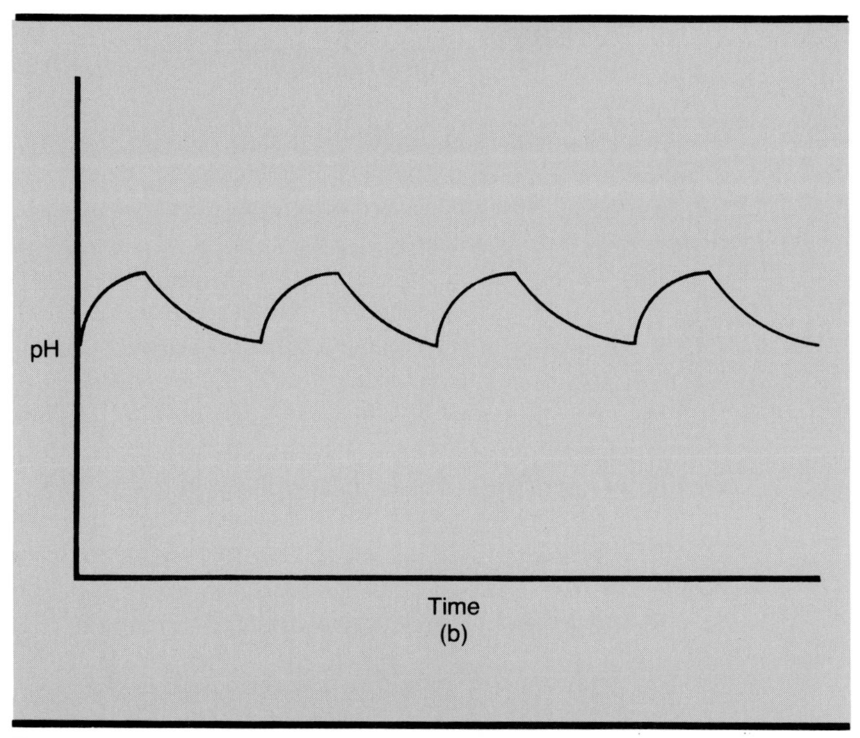

Fig. 6-3b. An overlap in the split range of an acid and base reagent valve causes a sawtooth pH oscillation.

Normally, the center of the signal range is used as the transition point for the split ranging of an acid or base valve. While this point is easy to remember for maintenance and operations, a better transition point for loop performance is one that equalizes the control valve gain seen by the controller over the whole signal range. The ratio of the portions of the signal range allocated for the acid and base should be equal to the ratio of the product of reagent valve capacity and reagent normality for the acid and base, as shown by Eq. (6-4). The same type of relationship also holds for the split ranging of different or identical normality acids or bases (the normalities cancel out when equal).

Thus, the transition point for a large linear trim valve whose capacity is ten times that of a small linear trim valve on the same reagent should be at 10%. The small valve would stroke from 0 to 10% and the large valve would stroke from 10 to 100% of the signal range.

$$S_1 = \frac{F_{1max} * N_1}{F_{2max} * N_2} * S_2 \qquad (6\text{-}4a)$$

$$S_1 = 100 - S_2 \qquad (6\text{-}4b)$$

where:

F_{1max} = maximum flow of reagent number 1 (gpm)

F_{2max} = maximum flow of reagent number 2 (gpm)

N_1 = normality of reagent number 1 (gm equivalents/liter)

N_2 = normality of reagent number 2 (gm equivalents/liter)

S_1 = portion of signal span allocated to reagent number 1 (%)

S_2 = portion of signal span allocated to reagent number 2 (%)

If an acid and base valve are split ranged, each has an opposite effect on the pH measurement. Since only a single pH control action can be used, one of the valve actions has to be the reverse of the other. If the two reagent valves have the same failure modes, and thus the same actions, a signal reversing relay must be used for one of the valves. Some valve positioners and I/P transducers have an optional signal reversing relay internally. For a reverse pH control action, the acid valve signal should be reversed and for a direct pH control action, the base valve signal should be reversed if the control valves both fail closed. The opposite valve signal assignment should be used if the control valves are both fail open.

Various computations have been developed for split ranging equal percentage characteristic valves. While the nonlinear gain can for some specific set of influent flow and concentration conditions partially compensate for the nonlinearity of the titration curve, the assumption that these conditions always exist is tenuous. The use of reset action in a pH controller means that there is no one-to-one relationship between the measurement signal and the control valve signal. The pH controller will drive its output to whatever value is necessary to reduce the error. Thus, nonlinearity in the measurement signal cannot generally be compensated by control valve trim nonlinearity for proportional-integral (PI) or proportional-integral-derivative (PID) controllers. It also cannot be used for feedforward pH control if a PI or PID controller is used to correct the valve position for feedforward calculation error. Reagent control valves with linear installed characteristics are recommended to avoid the addition of another nonlinearity to a difficult loop. If the valve pressure drop and flow is a significant portion of the piping system pressure drop and flow in a pumped system, an inherent equal percentage characteristic will distort to a linear installed characteristic and a linear inherent characteristic will distort toward a quick-opening installed characteristic. However, the reagent flow is normally extremely small and taken off a header supplying other users so that the inherent and installed characteristics are equal. For those exceptions where the flow characteristic does distort, the installed rangeability will be reduced from the inherent rangeability for liquid flow per Eq. (6-5a) (Ref. 3). The inherent rangeability can be approximated as the inverse of the fractional valve hysteresis for those valves with a linear flow characteristic that does not become erratic below the minimum lift set by hysteresis, as shown by Eq. (6-5b).

$$R_f = R_i * \sqrt{\frac{\Delta P_1}{\Delta P_2}} \tag{6-5a}$$

for well-defined linear flow characteristic:

$$R_i = \frac{100}{E_v} \tag{6-5b}$$

where:

ΔP_1 = pressure drop at maximum flow (psi)

ΔP_2 = pressure drop at minimum flow (psi)

E_v = control valve hysteresis (%)

R_i = inherent (shelf) rangeability

R_f = installed (field) rangeability

A small reagent control valve in parallel with a large reagent control valve will solve a reagent delivery precision and rangeability problem, but the two valves will interact. To reduce this interaction problem, the output of a separate proportional-only controller can be used to position the small valve, as shown in Fig. 6-4a, or the output of a valve position controller can be used to position the big valve, as shown in Fig. 6-4b. The measurement for the valve position controller is small valve position and the setpoint is 50%. The valve position controller has the advantage of insuring that the small valve precision is always available by keeping it near the middle of its stroke range by slowly adjusting the big valve. Since the valve position controller has to be detuned (proportional band increased) to reduce interaction between it and the pH controller, this configuration is sluggish for large load changes.

Metering or positive displacement pumps with pneumatically set stroke lengths typically have a precision of 2 to 5% and a drop off in flow at 5 to 10% of capacity (10:1 to 20:1 rangeability). The accuracy of the pneumatic calibration also deteriorates with time. Motor-driven metering pumps with electronically set stroke frequencies have better precision but about the same rangeability. Electronic metering pumps that use a DC solenoid and solid-state pulsing circuit instead of a conventional motor and gearbox assembly have a rangeability of 30:1 by a combination of a stroke frequency and length adjustment. The pulsing flow and pressure from a metering pump creates reagent injection problems and pH measurement noise problems for inline pH control systems.

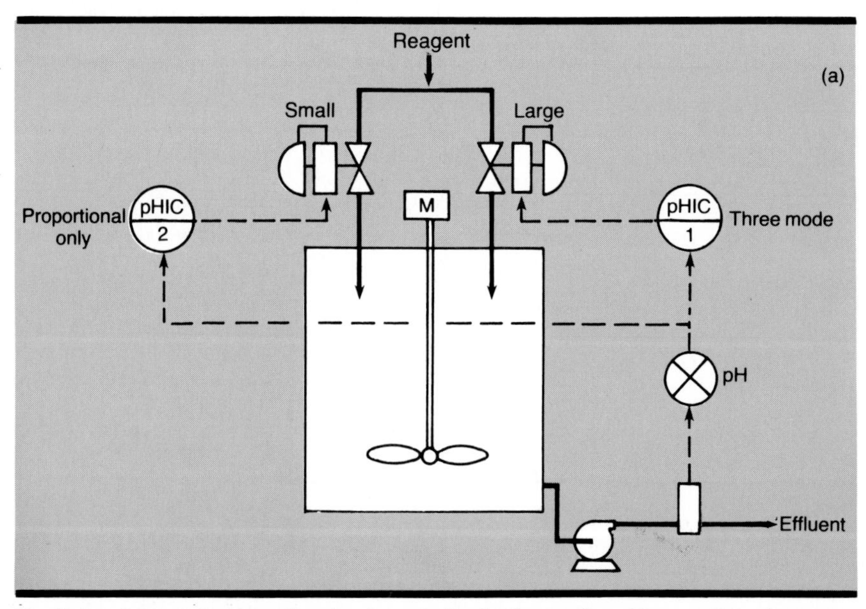

Fig. 6-4a. A separate proportional-only controller for the smaller of the parallel valves can be used to achieve reagent precision and rangeability requirements with minimal interaction.

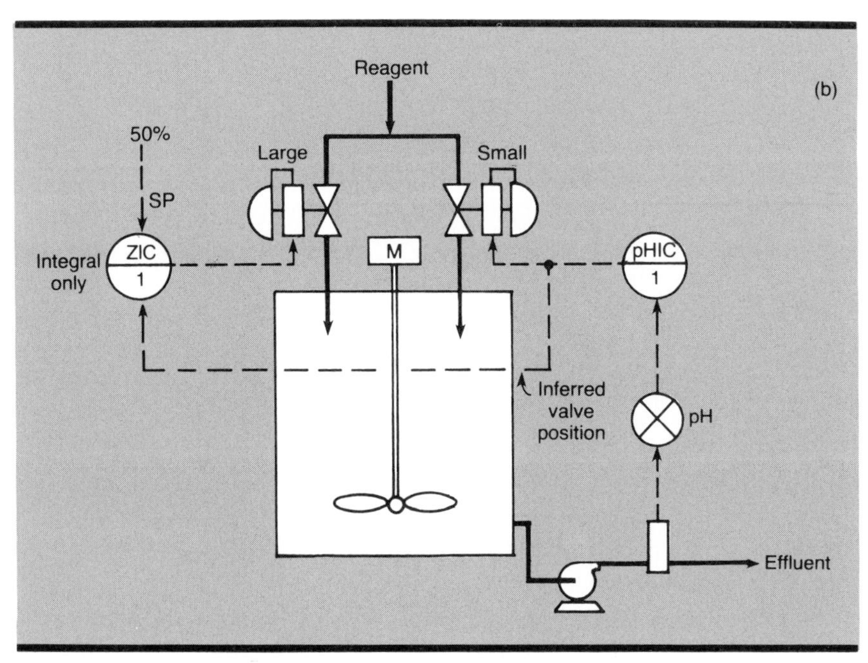

Fig. 6-4b. A valve position controller to slowly adjust the bigger of parallel valves to maintain the smaller valve near mid-range can be used to achieve reagent precision and rangeability requirements with minimal interaction.

6-3. Dynamics

The actuators for reagent control valves are usually small since the capacity required, and hence the size required, is usually small for pH control. Also, control valve hysteresis that increases the prestroke dead time for slowly changing signals is minimized to meet the reagent delivery precision requirement. Consequently, the reagent control valve prestroke dead time and stroking time contribute a negligible amount to the loop dead time except for a small, well-mixed vertical tank or static mixer with pipeline injector electrodes close to the effluent discharge.

If the controller output for such fast loops is changing at a rate faster than the velocity-limited exponential response of the control valve stroke will allow, the velocity limit will contribute a lag. The increase in the prestroke dead time for a change in stroke direction is proportional to the hysteresis and inversely proportional to the rate of change of the controller output signal (Ref. 4).

Thus, the velocity-limiting effect is more important for fast loops and the hysteresis effect is more important for slow loops. Of course, a high-friction packing on a valve without a positioner or a rotary valve will cause a large prestroke dead time for fast loops as well.

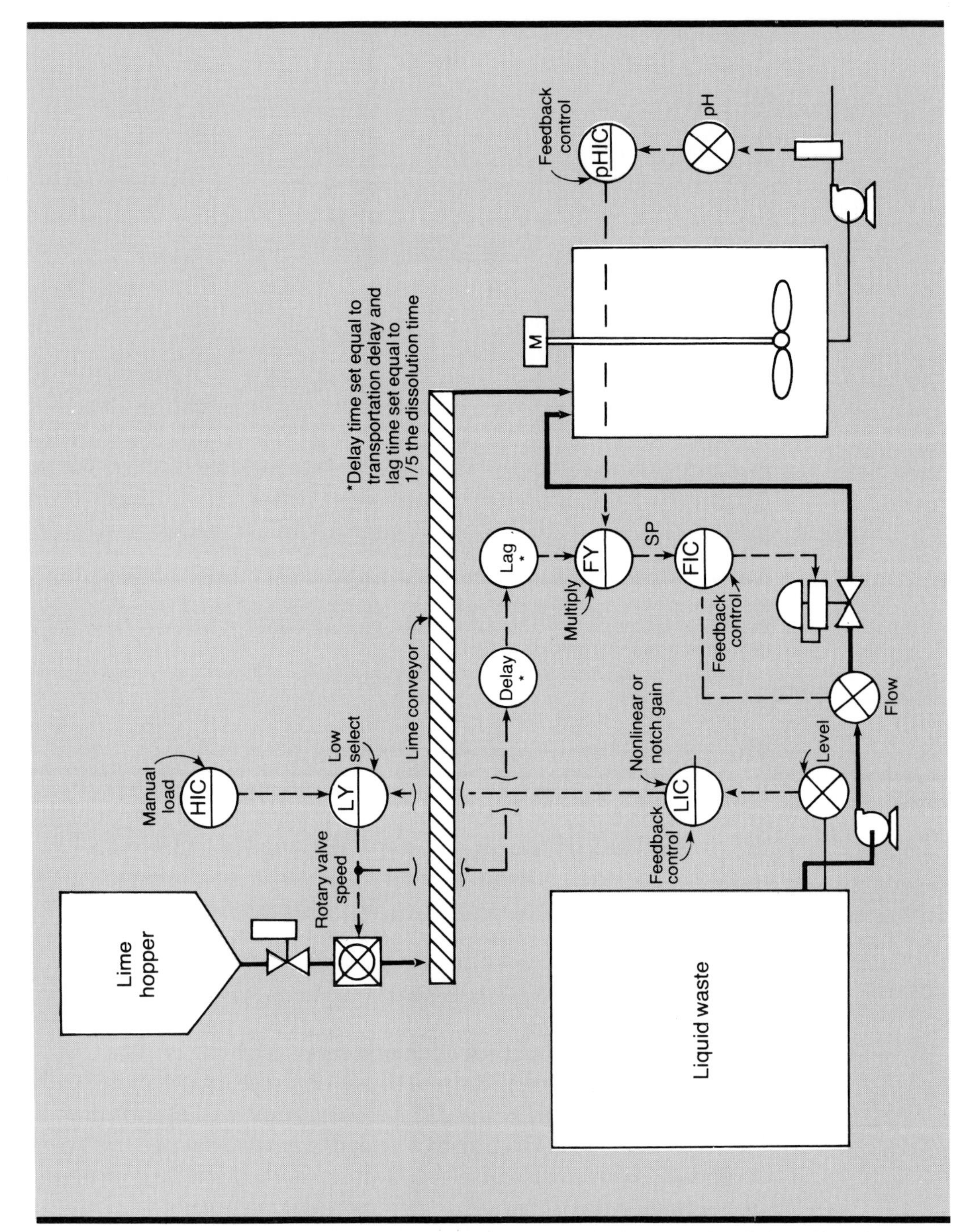

Fig. 6-5. The reagent transportation delay of a lime feeder can be eliminated from the pH control loop by base loading the lime feeder speed and throttling the influent flow.

Lime feeders have a transportation delay that is proportional to the length of the feeder divided by its speed. This transportation delay may be several minutes. The residence time of a vertical tank must be at least twenty times as large as the sum of this delay plus the equipment dead time for the vessel to be classified as well-mixed. An alternative to increasing the vessel size, and correspondingly the agitation power, is to base load the lime feeder speed and have the pH controller throttle the influent flow. The level controller on the influent tank slowly corrects the lime feeder base speed if the level gets too high or low. A non-linear or notch gain level controller is used to suppress control action when the level is in mid-range. A 50% output from the level controller corresponds to a zero correction of the base-loaded speed. The operator can change the system capacity by changing the base loaded speed through a manual loader. Figure 6-5 shows this method of eliminating the reagent transportation delay of the lime feeder from the pH control loop.

6-4. Types

Reagent valves differ from other control valves by requiring a lower capacity and lower hysteresis. The Masoneilan MicroPak, Baumann 24000 (765) series, and Fisher 585-CE (3611) series have low-capacity trims and high-performance positioner actuator assemblies that reduce the hysteresis to less than 1%. The dead band of the I/P transducer should be checked to make sure that it is less than that of the high-performance positioner. Masoneilan also has a VariPak valve that is a MicroPak valve with an adjustable maximum-flow coefficient that can be set to pass the maximum flow identified after startup. Since the hysteresis remains a fixed percent of the chosen stroke, hysteresis as a percent of capacity is decreased by the elimination of excess capacity.

If the reagent viscosity or flow is low enough to have a Reynolds number below 5000 for the diameter of the orifice in the control valve, the flow pattern will change from turbulent to laminar. The transition from turbulent to laminar flow in a conventional control valve causes a large change in the actual flow coefficient, and hence the valve gain and rangeability, because flow is proportional to the pressure drop to the one half power for turbulent flow and to the first power for laminar flow.

The laminar flow valve developed by Hans D. Baumann provides tremendous inherent rangeability for small flows (Cv < 0.01) because the flow characteristic varies with the third power of the stroke as shown in Fig. 6-6. For a stroke rangeability of 25:1, the inherent rangeability is 15000:1. Since the change in pump head and system frictional pressure drop are negligible for these small flows, the installed rangeability is equal to the inherent rangeability (Ref. 5). The control valve gain is proportional to the second power of the stroke but this nonlinearity can be easily compensated for by the use of a ⅓ fractional power functional

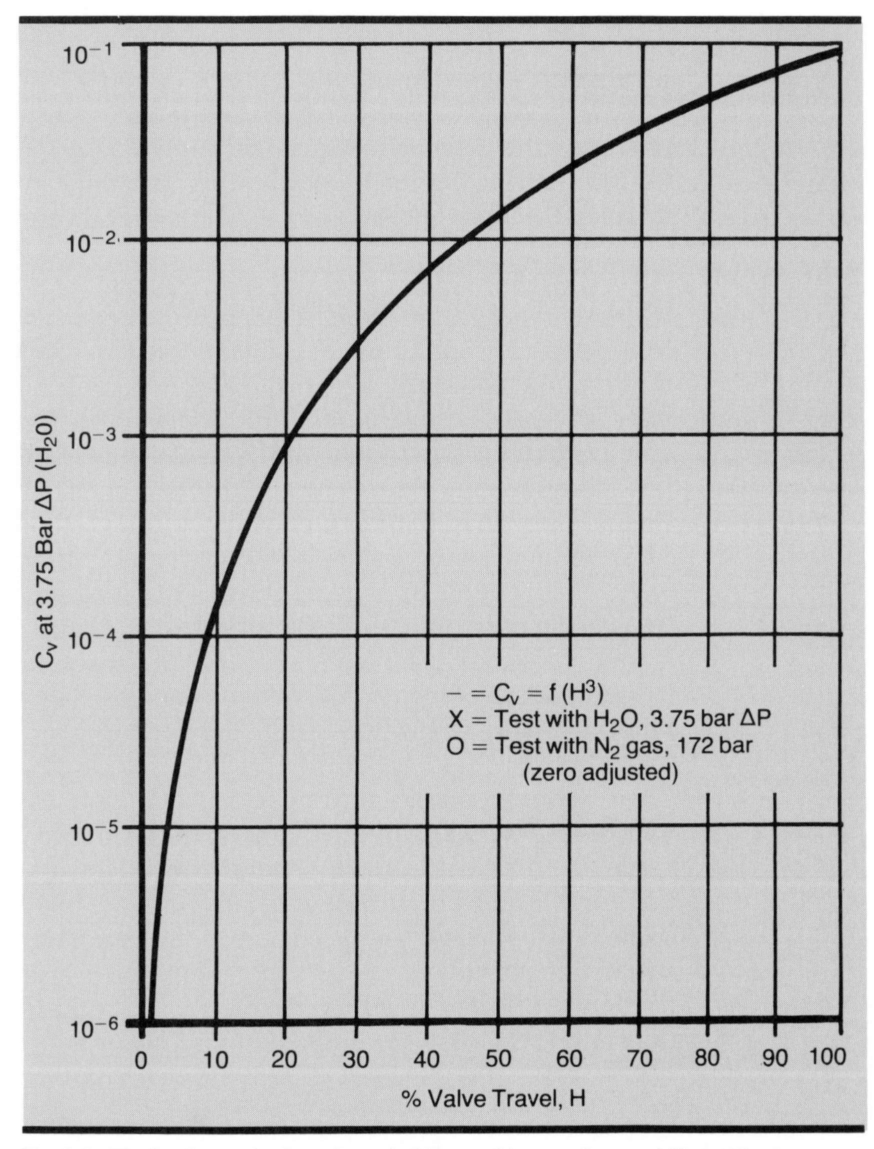

Fig. 6-6. The laminar valve flow characteristic provides great rangeability but its slope varies, which means the valve gain is nonlinear.

block in a microprocessor on the controller output. An analog square root extractor could be used between the controller and the valve with some reduction in valve rangeability due to signal inaccuracy and dropout at low signal levels.

The digital valve developed by Process Systems, Inc. (a division of Powell Industries, Inc.) provides an inherent rangeability determined by the number of binary coded ports per Eq. (6-6). The valve stroke hysteresis is equal to the inverse of the valve rangeability multiplied by 100%. Thus, the inherent rangeability is 4096:1 and the stroke hysteresis is about 0.02% for a 12 bit digital valve. The lives of the individual solenoids are prolonged by reducing the power input periodically (the duration of the reduction in power is not long enough to switch the valve), which reduces the operating temperature of the solenoids. The life of the smallest port and solenoid is prolonged by the use of a filter on the valve signal to prevent dithering. The inherent flow characteristic is linear, the sum of the prestroke dead time and stroking time varies from 25 to 100 milliseconds, and the stroke overshoot is zero. These features are impressive, but there are some important problems to be reckoned with in the maintenance of the digital valve. Since the digital valve uses soft seats, it is susceptible to erosion damage by particulates or flashing. The sticking or failure of an individual actuator cannot be determined by external inspection. The smallest port is susceptible to plugging in fouling service since it is the most frequently used port.

$$R_i = 2^n \tag{6-6}$$

where:

n = number of binary coded ports

R_i = inherent rangeability

If the flow is very low, the control valve orifice size required for continuous throttling is so small that it becomes unreliable due to excessive plugging. While upstream strainers can be used to filter out some particulates, the suspension in sulfuric acid or the oil in gaseous sulfur dioxide will plug both the strainer and the control valve. Also, the inherent characteristic becomes more erratic since the dimensional error of the trim as a fraction of the flow path clearance increases as the trim size decreases for a given manufacturing tolerance. Pulse amplitude modulation, pulse width modulation, or pulse interval modulation can be used to provide a larger orifice for flow since the valve is periodically pulsed instead of continuously throttled. Pulse amplitude modulation and pulse width modulation provide a linear valve gain, but the amplitude or pulse width may become so large in an attempt to achieve sufficient rangeability that the pulse shows up as extensive noise in the pH measurement. The effect of a given reagent pulse amplitude and duration on the pH measurement can be found by taking the amplitude of the filter reagent pulse amplitude from Eq. (6-7) and translating that graphically on a titration curve to a change in pH measurement at the pH setpoint. The inherent rangeability of pulse amplitude modulation is approximately equal to the maximum pulse amplitude divided by the minimum pulse amplitude. The installed rangeability depends upon the change in pressure drop with the change in pulse amplitude. The minimum pulse amplitude must be large enough to exceed the stroke hysteresis and to sufficiently reduce the susceptibility of the valve to plugging.

The maximum pulse amplitude is set by Eq. (6-7). The installed rangeability of pulse width modulation is approximately equal to the maximum pulse width divided by the minimum pulse width. The minimum pulse width must be larger than the sum of the valve prestroke dead time and stroking time. The maximum pulse width is set by Eq. (6-7). Pulse width modulation can be implemented by the addition of a Moore Industries PDT pulse on the output of any pH controller or by the use of the DAT option in the Leeds & Northrup 7084 series microprocessor pH controller. Pulse interval modulation (pulse frequency modulation) is less likely to cause pH measurement noise because the pulse amplitude and width are kept at the minimum, but it has a nonlinear valve gain. The valve gain is very low for low controller outputs if the maximum interval selected is very large. This nonlinearity can be compensated for by multiplying the controller output by the inverse of the gain documented in Ref. 5 for pulse interval control. However, the complexity of the calculation is greater than that described for the laminar flow valve. Figure 6-7 shows the flow characteristic for a selected pulse width and maximum pulse interval. The installed rangeability of pulse interval control is approximately equal to the maximum pulse interval divided by the pulse width. The maximum pulse interval should not be larger than twice the turnover time of the tank so that the responsiveness of the loop is not hindered. The minimum pulse amplitude and width should be selected per the aforementioned criteria for pulse amplitude and width modulation. Pulse interval control can be implemented by installation of a Rochester SC-1356 modified current to pulse frequency converted on the output of any pH controller.

for $T_p < TC_p$:

$$A_f = \frac{A_p * T_p}{TC_p * \pi} \quad (6\text{-}7)$$

where:

A_f = filtered reagent pulse amplitude (pph)

A_p = reagent pulse amplitude (pph)

T_p = reagent pulse width (minutes)

TC_p = vessel equipment time constant (minutes)

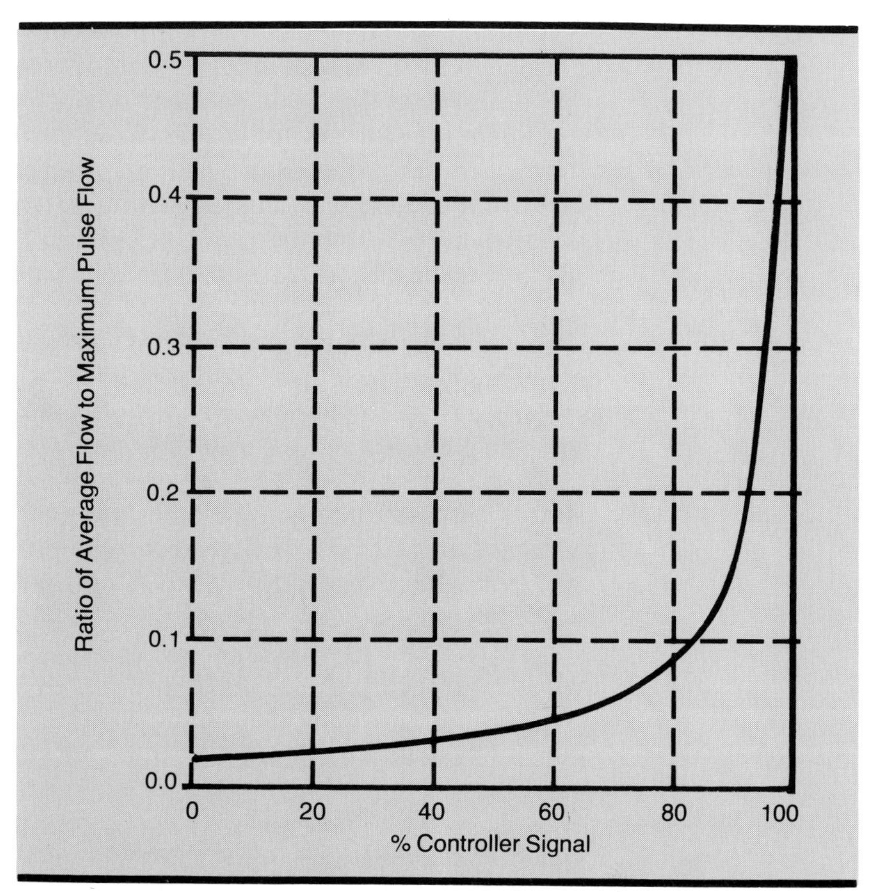

Fig. 6-7. Pulse interval control causes less noise in the pH measurement from pulses but the slope of the flow characteristic varies, which means the valve gain is nonlinear.

A globe control valve that is pulsed must be rugged enough to withstand the repeated hammering of the valve seat by the valve plug. A ¼ or ½ inch piston-actuated ball valve is usually less expensive to replace and more able to withstand on-off service than a globe control valve. Thus, a more practical arrangement for pulse width or interval control is to pulse a small ball valve downstream of a globe control valve that is manually positioned to set the pulse amplitude. Even so, the life of the pulsed ball valve may not exceed six months. An installed spare ball valve in a parallel reagent line should be provided to minimize downtime.

Exercises

6-1. *What is the maximum allowable hysteresis for each of the titration curves shown in Figs. 6-1a and 6-1b?*

6-2. *If the capacity of a control valve is doubled due to lack of a trim size in the model number preferred for spare parts, what happens to the pH error due to reagent error for a titration curve slope equal to 10?*

6-3. *How could you detect a leaking acid reagent control valve from the control room for a split-ranged acid and base reagent system if there were no reagent flow measurements?*

6-4. *What is the transition point between split-ranged air fail closed acid and base reagent valves if the acid valve normality is 10, the base valve normality is 2.5, and the capacity of the valves are both 100 pph?*

6-5. *What is the prestroke dead time due to hysteresis for a 3-15 psi signal range, TFE-asbestos packing, globe valve, no positioner, and a controller output signal changing at the rate of 1% per second?*

References

[1]McMillan, G. K., *Tuning and Control Loop Performance*, ISA Monograph 4, 1983, pp. 142-143.
[2]Schuder, C. B., "Control Valve Rangeability and the Use of Valve Positioners," *Advances in Instrumentation*, ISA 26th Annual Conference, Vol. 26, Part 4, Paper 817, Oct. 1971, pp. 2-3.
[3]Shinskey, F. G., *Energy Conservation through Control*, Academic Press, 1978, pp. 32-33.
[4]Baumann, H. D., "Control Valve with Laminar Flow Characteristic," ISA Conference, Paper 540, March 1981, pp. 179-186.
[5]McMillan, G. K., *Tuning and Control Loop Performance*, ISA Monograph 4, 1983, pp. 150-151.

Unit 7:
pH Reagent

UNIT 7

pH Reagent

This unit describes the beneficial and detrimental effects of certain reagent characteristics and piping installations on system performance.

Learning Objectives—**When you have completed this unit you should:**

A. **Know what is to be gained by reagent dilution.**

B. **Know what is to be gained by reagent buffering.**

C. **Be able to estimate the detrimental effect of dissolution time on system performance.**

D. **Be able to avoid reagent piping installation pitfalls.**

7-1. Dilution

A common misconception is that the slope of the titration curve, and hence the sensitivity of the system, can be decreased by reagent dilution. Reagent dilution has a negligible effect on the shape of the titration curve: the curve slope will appear larger if the same abscissa is used because only a portion of the original curve is displayed. The numbers along the abscissa must be multiplied by the ratio of the old to new reagent concentration to show the entire original titration curve. For example, the abscissa values would have to be doubled if the concentration of the reagent were cut in half. While the allowable reagent error band is increased, the reagent control valve capacity requirement is increased by the same factor so that the valve hysteresis requirement does not change. Reagent dilution is beneficial if the reagent flow required is so small that reagent valve-plugging frequency, reagent transportation delay, or reagent viscosity is too great. The freezing point and winterization problem is decreased for sodium hydroxide by dilution. The plugging tendency of sulfur dioxide is significantly reduced by the addition of water. While some reagents, such as sodium hydroxide, become less corrosive to steels when diluted, other reagents, such as sulfuric acid, become more corrosive to steels when diluted.

If reagent dilution is used, the system must be carefully designed to prevent the creation of reagent concentration upsets and reagent delivery delays. The pH controller should throttle the diluted reagent and the water flow should be ratioed to the reagent flow. If the pH controller throttled the undiluted reagent or the water, the reagent delivery time delay is for the plug flow concentration response instead of for a liquid flow response. The time delay for the concentration response is approximately equal to the residence time whereas the time delay for a full pipeline liquid flow response is less than 0.1 second. If the titration curve is steep, the fast concentration disturbances from flow measurement inaccuracy and noise may cause pH excursions outside of the control band. For example, the single step neutralization of 0 pH hydrochloric acid with sodium hydroxide would require a flow measurement accuracy of 0.0001% for reagent dilution to achieve a control band of 2 pH about a 7 pH setpoint. If a static mixer is used for pH control, the fast concentration disturbances from reagent dilution will create pH measurement noise. Therefore, for steep titration curves and inline pH control, a storage tank for diluted reagent should be installed to filter the fast reagent concentration disturbances from dilution.

7-2. Buffering

The buffering mechanism was described in Sec. 3-5. The addition of a buffer with a dissociation constant pH equal to the pH setpoint can greatly reduce the slope and hence the system sensitivity in the pH control band. The buffer capacity, also known as the buffer value, is defined as the change in normality due to the addition of a strong base or acid divided by the change in pH as shown in Eq. (7-1). The buffer capacity is a measure of the ability of the solution to resist a change in pH. Typical buffer capacities of buffer solutions used for pH meter calibration range from 0.01 to 0.2. The equations in Unit 2 can be used to convert from normality to weight fraction or mass flow ratio of reagent to influent to increase the utility of the buffer capacity for industrial system design.

$$Z = \frac{\Delta N}{\Delta pH} \tag{7-1}$$

where:
ΔN = change in acid or base normality (gram-equivalents/liter)
ΔpH = change in solution pH
Z = buffer capacity (gram-equivalents/liter/pH)

7-3. Dissolution

Dissolution is the act of dissolving. Whereas the pH reaction time rarely exceeds a few seconds, the dissolution time of some solid or gaseous reagents can be several minutes. The dissolution time is equivalent to the batch conversion time. It can be found by placing an influent sample in a beaker, adding a shot of reagent, and noting the time between the shot of the reagent and 99% conversion. The beaker must have the same degree of agitation (equipment dead time to time constant ratio) as the field vessel. A pH measurement can be used to note the 99% conversion point but a very steep titration curve slope will probably result in a longer than actual and a very flat titration curve will probably result in a shorter than actual dissolution time measurement. If the reagent is continuously added instead of pulsed, the pH measurement may suddenly approach the final value and then rapidly return back to the previous value due to dynamic changes in solution saturation. These "rebounds" add ambiguity to the time measurement.

A back-mixed vessel is much better than a plug flow vessel for control due to the much lower dead time to time constant ratio. However the residence time distribution has a much greater spread. Some of the material is in the vessel for a time much smaller than the residence time for continuous operation. Consequently, a greater residence time is required for a continuous back-mixed vessel for complete conversion. The plot in Fig. 7-1 shows that the residence time of a continuous back-mixed vessel must be twenty times the dissolution time for 99% conversion (Ref. 1). The agitator pumping rate must be scaled up in proportion to the vessel volume so that the equipment dead time to time ratio remains the same. Since the horsepower required is proportional to the third power of the pumping rate, the additional energy cost may be greater than the savings from using lime or gaseous ammonia instead of caustic as a reagent. The capital cost of the larger vessel and motor may also be significant. If the size of a single back-mixed vessel becomes prohibitive, several smaller ones in series can be used to achieve the desired conversion percentage. The effect of these vessels in series on conversion is multiplicative in terms of conversion fractions as shown by Eq. 7-2. For example, two vessels—each with a residence time that is twice the dissolution time—will have an individual conversion of 90% and an overall conversion of 99%.

$$X_0 = X_1 + (1 - X_1) * X_2 \qquad (7\text{-}2)$$

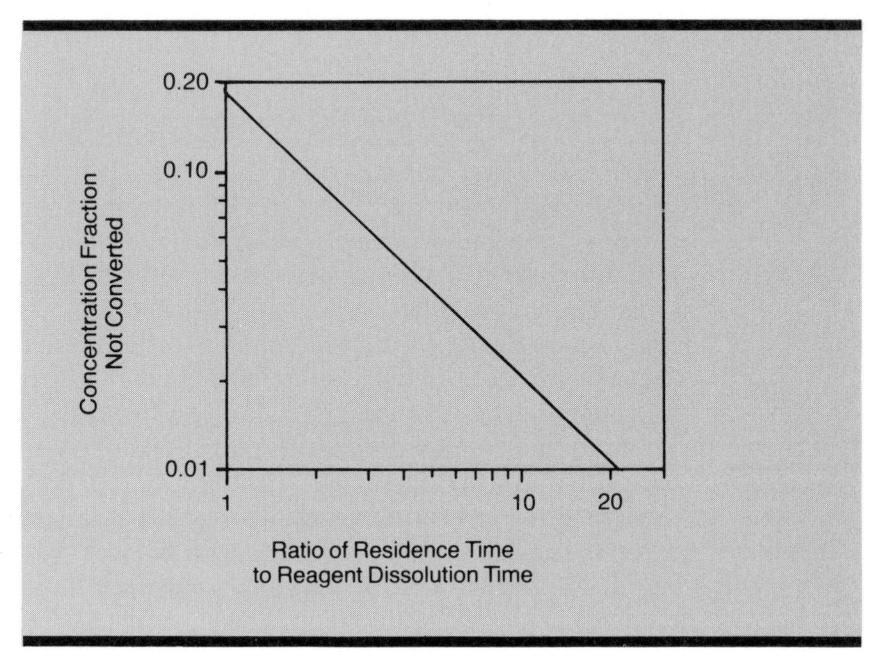

Fig. 7-1. A back-mixed continuous vessel requires a residence time many times the dissolution time to get complete conversion.

where:
X_0 = overall conversion fraction at outlet of last vessel

X_1 = conversion fraction in first vessel

X_2 = conversion fraction in second vessel

The dissolution time also has an effect equivalent to a slow reagent control valve because the change in the quantity of dissolved reagent lags the change in pH controller output. The equivalent slow valve time constant is approximately equal to one-fifth the dissolution time. Most of this time constant will be converted to loop dead time that will increase the ultimate period of the loop unless the time constant is negligible compared to the equipment dead time. Unit 9 will quantitatively show the effect of this time constant on the ultimate period and the controller mode settings. If the influent flow can be throttled, it is best to eliminate the slow valve effect by base loading the reagent and pH controlling the influent in the fashion described in Sec. 6-3 for lime feeders.

The dissolution time of *limestone* is too large to allow its use for feedback control even when pulverized into fine particles. The

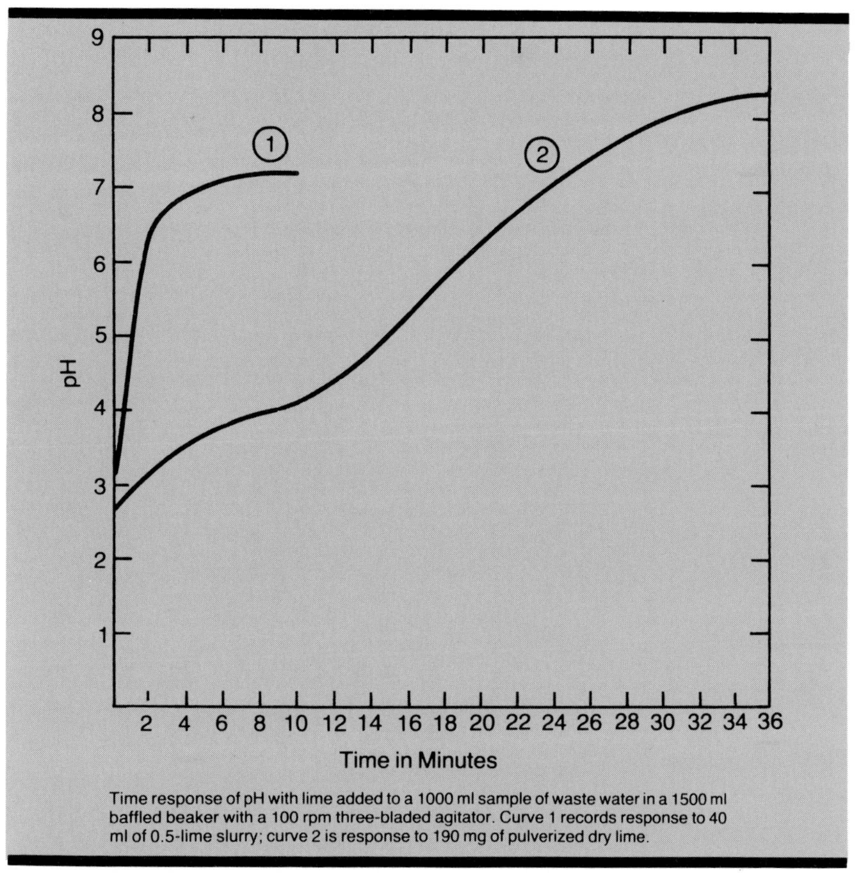

Time response of pH with lime added to a 1000 ml sample of waste water in a 1500 ml baffled beaker with a 100 rpm three-bladed agitator. Curve 1 records response to 40 ml of 0.5-lime slurry; curve 2 is response to 190 mg of pulverized dry lime.

Fig. 7-2. The dissolution time of pulverized dry lime is reduced by slaking the lime and making it into a slurry.

main use of limestone is for manual treatment of acid spills by covering the spill area or diverting the spill to a limestone pit. The dissolution time of pulverized *dry lime* can be greatly reduced by slaking the lime and making it into a *lime slurry*. Figure 7-2 shows that the dissolution time for pulverized dry lime is about 32 minutes and the dissolution time for lime slurry is about eight minutes when added to a 1000 milliliter sample of waste water in a 1500 milliliter baffled beaker with a 100 rpm three-blade agitator (Ref. 2). The above data is for "soft burned" lime. If the lime is heated in the kiln longer, the lime becomes "over-burned" or "head burned" and takes a longer time to dissolve (Ref. 3). Dry lime tends to lay on the surface unless wetted and forced below the surface by a water spray. Therefore a water spray should be used for dry lime addition (Ref. 4). The dissolution time of lime slurry increases with age due to the agglomeration of small particles into larger particles even though the lime slurry storage tank is mildly agitated (Ref. 5). If the pH controller manipulates the lime feeder or water addition to the lime slurry storage tank, the equipment time constant of the slurry tank has the same effect as a slow valve time constant. To prevent the addition of this time constant to the loop, the lime feeder speed should be manipulated by a slurry tank level controller and the water addition rate ratioed to the lime feeder speed. The pH controller should manipulate the lime slurry addition to the neutralization vessel. The lime slurry feed must be kept flowing by the use of a recycle loop to prevent settling and plugging of the reagent lines. A throttled globe control valve will plug. The liner of a pinch diaphragm valve will fail due to erosion. Pulse width or pulse interval modulation of an on-off ball valve is the best alternative because it has the least maintenance problems and is relatively inexpensive to periodically replace. Section 7-4 has information on how to design a pulse width and pulse interval modulation system.

Precipitation can force the pH of the effluent to a level that may be outside the control band. The precipitation of ferric and ferrous hydroxides acts as a reservoir of hydroxyl ions. An 11 pH upset for an hour will increase the reservoir of hydroxyl ions so much that the pH may remain in the 9 to 10 range for a day or more (Ref. 6). A given salt will precipitate when the product of the ion concentrations exceeds the solubility product constant. The relative solubility of different salts can be compared by dividing the pK for the solubility product constant by the sum of the coefficients of the ion concentrations (Ref. 7). The relative

Acid Neutralized by Lime	Salt Precipitated by Lime	Solubility Product pK	Sum of Coefficient	Relative Solubility
Sulfuric acid	Calcium sulphate	4.2	2	2.1
Hydrofluoric acid	Calcium flouride	10.4	3	3.4
Carbonic acid	Calcium carbonate	8.1	2	4.1
Phosphoric acid	Calcium phosphate	26.	5	5.2

Table 7-1. The tendency to precipitate increases as the relative solubility increases.

solubility of salts from the neutralization of various acids by lime (calcium hydroxide) is shown in Table 7-1. A high number indicates a greater tendency to precipitate.

7-4. Piping

The pipe length between the reagent control valve and the reagent injection point should be vertical and as short as possible. The dynamic response of full pipeline liquid flow is very fast with a time constant of about 0.1 second and a dead time much less than 0.1 second. Liquid flow at the exit of a full pipeline starts when the incompressible column of fluid starts to accelerate from a pressure wave traveling at the speed of sound initiated by the change in position of the reagent valve (Ref. 8). The flow rate of most reagents is so small that the pipeline is only partially full and may be empty in spots. When the reagent control valve opens, the flow does not increase at the injection point until the increment in flow travels the distance between the valve and the injection point. When the control valve closes, the flow does not stop at the injection point until the pipeline empties. Therefore, a transportation delay is created that is equal to the pipeline length divided by the velocity. The velocity of the reagent in a partially wetted vertical pipeline can be approximated as the velocity of a falling film by Eq. (7-3). For water and a film thickness of 0.01 feet, the falling film velocity is about 1 fps. Reagent dilution can be used to create full pipeline flow and reduce the reagent delivery time delay.

$$v = \frac{z^3 * d * g}{3 * u} \qquad (7\text{-}3)$$

where:

d = density of the reagent (lb/cu ft)

g = acceleration due to gravity $(\mathrm{ft/sec})^2$

u = viscosity of the reagent (lb/ft*sec)

v = velocity of falling film (fps)

z = thickness of falling film (ft)

An even greater time delay develops when process fluid from the vessel or static mixer back fills the reagent pipeline after the control valve closes. The reagent must displace this volume of fluid. What may normally seem as negligible lengths of reagent pipe can create a large time delay due to extremely low reagent flow rates. This time delay is approximately equal to the volume of the reagent pipe from the injection point to the level surface for a vessel and to the reagent control valve for a static mixer divided by the reagent volumetric flow as shown by Eq. (7-4). For example, a 5-foot section of 1-inch diameter schedule 40 pipe and a reagent flow rate of 2 gph will create a time delay of 6 minutes.

For a liquid reagent pipeline back filled with process fluid:

$$TD_v = \frac{V_r}{Q_r} \tag{7-4}$$

where:

Q_r = reagent volumetric flow (gpm)

TD_v = time delay to start flow when reagent valve reopens (min)

V_r = volume of reagent pipeline that is back filled (gal)

For a pipeline, the reagent injection velocity should be at least twice the influent velocity to get some gross mixing of the reagent and influent before entering the static mixer. Also, the reagent injection Reynolds number should be greater than 5000 to avoid the formation of a laminar reagent jet. For a Reynolds number between 200 and 800, the laminar jet is 18 injection nozzle diameters long. It is difficult to get a sufficient injection velocity and Reynolds number for sulfuric acid because its suspension requires a large injection hole to prevent plugging and because its high viscosity decreases the Reynolds number. If the injection hole diameter cannot be increased, the velocity of the influent at the injection point can be decreased by swagging up the pipeline. Better premixing of the reagent and influent can be obtained by the use of multiple injection holes in a radial pattern instead of a single center injection hole. For example, a single center injection will require 125 equivalent pipe diameters for mixing whereas four radial injections will require 60 equivalent pipe diameters for mixing for a Reynolds number of 10,000 and a concentration variation of 1% (Ref. 9). Poor reagent injection will show up as pH measurement noise since there is relatively little axial or back mixing in an inline system.

For a vessel, the reagent injection point should be just below the surface near the agitator shaft so that the reagent is pulled down into the eye of the impeller by the axial agitation flow pattern. The injection point should be as far as possible from the exit nozzle to prevent short circuiting to the effluent. If the exit nozzle is near the top of the vessel to regulate level by overflow, the length of the injection piping below the surface should be increased to about halfway down the agitator shaft as shown in Fig. 7-3. However, this will increase the time delay due to back filling of the reagent line. If the reagent is a gas, a sparger that extends down to the impeller should be used to disperse the gas near the bottom so that more gas will dissolve before the rising bubbles reach the surface.

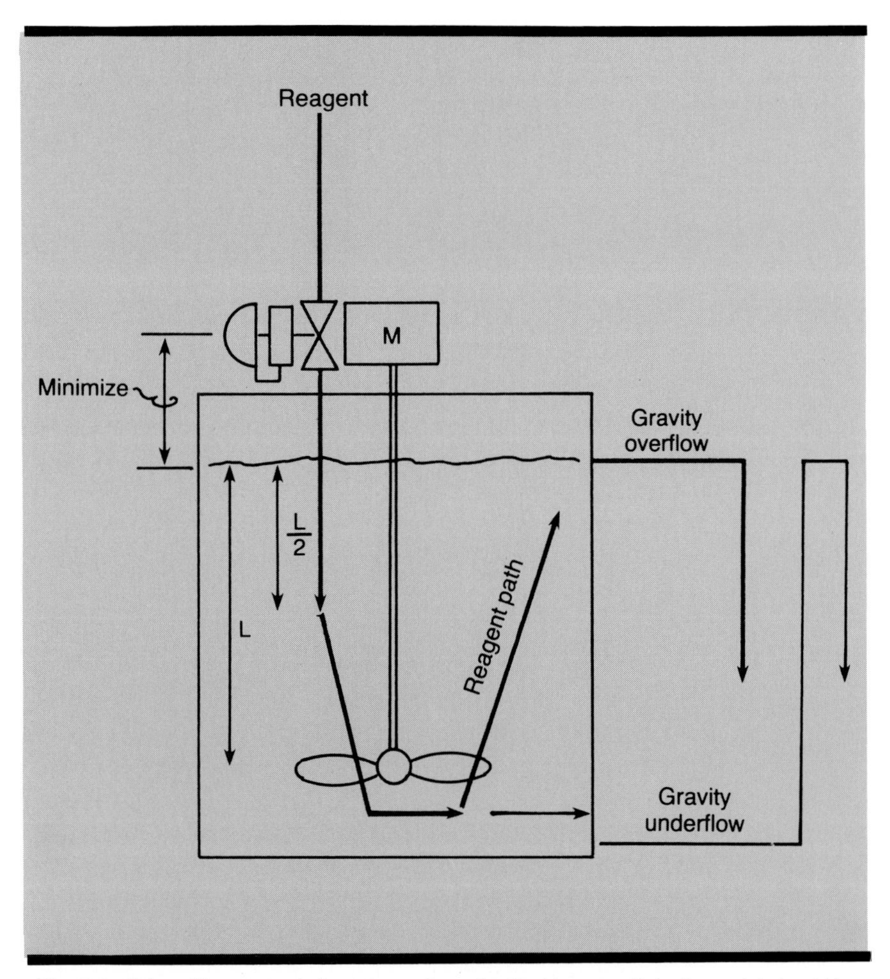

Fig. 7-3. If the effluent nozzle is at the surface, the liquid reagent injection point should be halfway down the agitator shaft. For gravity underflow, the injection point should be just below the surface.

If the reagent header is not dedicated to the pH system but has multiple users, a pressure regulator should be installed upstream of the reagent control valve to minimize reagent errors due to reagent pressure disturbances. If flashing occurs in the reagent control valve during summer operation due to a high pressure drop across the valve, a pressure regulator should be installed upstream of the valve to absorb some of the pressure drop. The pressure drop across the pressure regulator should be less than that across the reagent control valve to prevent interaction between the pH and pressure loop.

Exercises

7-1. *List the advantages of a properly installed caustic reagent dilution.*

7-2. *Why would a test solution with a high buffer capacity be desirable pH meter calibration?*

7-3. *What is the residence time required for a 99% conversion in a single vessel for the dry lime dissolution time shown in Fig. 7-2?*

7-4. *If the reagent injection length below the vessel surface cannot be decreased, what other options exist to reduce the reagent delivery time delay due to back filling of the reagent pipe with vessel fluid?*

References

[1]Levenspiel, O., *Chemical Reaction Engineering*, John Wiley & Sons, 3rd edition, 1962, pp. 127-128.
[2]Moore, R. L., *Neutralization of Waste Water by pH Control*, ISA Monograph 1, 1978, pp. 89-90.
[3]*Ibid*, pp. 95-96.
[4]Shinskey, F. G., *pH and pION Control in Process and Waste Streams*, John Wiley & Sons, 1973, pp. 177-178.
[5]Moore, R. L., *Neutralization of Waste Water by pH Control*, ISA Monograph 1, 1978, pp. 93-94.
[6]Shinskey, F. G., *pH and pION Control in Process and Waste Streams*, John Wiley & Sons, 1973, pp. 73-74.
[7]*Ibid*, pp. 75-76.
[8]McMillan, G. K., *Tuning and Control Loop Performance*, ISA Monograph 4, 1983, p. 67.
[9]Clayton, C. G., et. al., "Dispersion and Mixing during Turbulent Flow of Water in a Circular Pipe," U.K. Atomic Energy Authority, AERE-R5569, 1968, pp. 1-5.

Unit 8:
pH Control Systems

UNIT 8

pH Control Systems

This unit describes the major types of pH control systems and the relative advantages and disadvantages of each.

Learning Objectives—When you have completed this unit you should:

A. Understand the basic differences between the different types of control systems.

B. Be aware of the recent technical developments to improve feedback and feedforward control.

C. Know how to utilize the power of the microprocessor-based controller to implement these improvements.

8-1. Feedback Control

Nearly all pH control loops utilize some form of feedback control as shown in Fig. 8-1. The reagent flow starts to correct for a disturbance originating from the influent or reagent only after it appears in the effluent, the controller reacts, and the reagent flow responds. The time between the start of the disturbance and the start of the reagent flow injection is equal to the total loop dead time. Thus, the peak pH error is reached after one loop dead time. The loop dead time is equal to the sum of the equipment time delay, the sample transportation delay, some fraction of the electrode time constant, some fraction of the digital measurement filter time constant, the reagent valve prestroke dead time, some fraction of the reagent valve stroking time, the reagent transportation delay, and some fraction of the reagent dissolution time. Thus there are many possible sources of delayed feedback correction and consequently poor feedback loop performance. Unit 9-1 will show how to estimate what fraction of a time constant is converted to dead time.

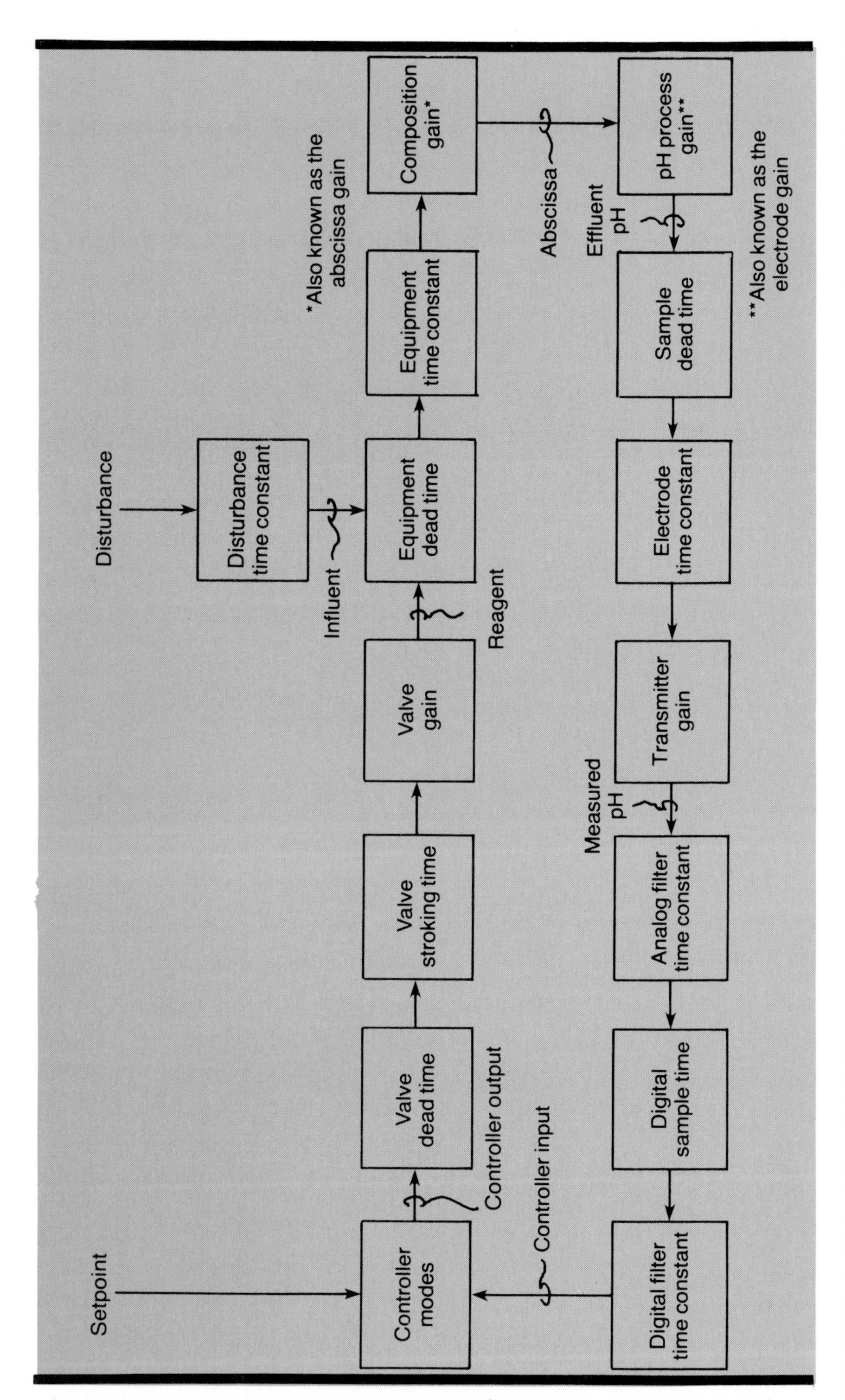

Fig. 8-1. The feedback control loop cannot correct for a disturbance until after one loop dead time which is the sum of all the pure time delays and the equivalent dead times from time constants.

8-2. Mode Characterization

Even if the total loop dead time is small, the action of a conventional feedback controller will be sluggish when the measurement is on the flat portion of the titration curve since the controller proportional band was increased (controller gain was decreased) for the steep portion of the titration curve. In order for a feedback loop to be stable, the product of the controller, valve, equipment, process, and measurement steady-state gains must be less than one at the ultimate period of the system. Thus the controller gain must be decreased to maintain stability for the high pH process gain for the steep portion of the titration curve. (Note that Shinskey and others consider the output of the process to be hydrogen ion normality so that the pH gain of the titration curve is referred to as an electrode gain instead of a process gain).

Characterization of the feedback controller gain mode can help achieve uniformity of the gain product and controller responsiveness. Nonlinear or notch gain controllers were developed more than a decade ago to provide two gain regions. The lower gain region is called the dead band or notch gain width and is set to match the steep portion of the titration curve. The start and end of the dead band or notch gain region is called the lower and upper breakpoints, respectively. These breakpoints are set as a percent of the pH measurement range except for one manufacturer where they are a percent deviation below and above the pH setpoint. If the breakpoints are based on the setpoint, they may have to be readjusted when the setpoint is moved. The controller gain reverts back to normal below the lower breakpoint and above the upper breakpoint. The breakpoint and dead band or notch gain settings are determined by making straight line approximations to the S-shape of the titration curve. A straight line is drawn parallel to the steepest portion, another line is drawn parallel to the tail of the curve that is flattest, and a final line is drawn with the same slope as the last line but parallel to the other tail of the curve. A fourth line is drawn from the pH for 0% transmitter output to the pH for 100% transmitter output. The notch gain is the slope of this line divided by the slope of the line parallel to the steep portion of the titration curve. Note that since this gain calculation is made from a ratio, the units of the abscissa of the titration curve are unimportant. The breakpoints are the pH intersection points of the three lines. The nonlinear and notch gain controller were developed based on the assumption that a titration curve is a single symmetrical S-shaped curve that can be approximated by straight lines. Actually, a titration curve can have multiple nonsymmetrical curves since a flat portion will occur at each major dissociation constant as explained in Unit 3. Also, the slope of the titration curve is continuously changing so that the straight line approximation is the result of a graphical deception. The resolution of a graph is insufficient to show the change in slope. In fact, the control band cannot be translated to a reagent error band for a strong acid and strong base curve. The slope of such a curve changes by a factor of 10 for each pH unit deviation from 7 pH as shown by Eq. (8-1) (Ref. 1). While the straight line approximation may look valid for a nonlinear controller as shown by Fig. 8-2a, a blowup of the center vertical section reveals another S-shaped curve and the misfit of the straight line approximation for the notch gain as shown by Fig. 8-2b. The straight line approximation becomes more valid as the degree of buffering increases.

For a strong acid and strong base:

$$\frac{\Delta pH}{\Delta N} = \frac{-0.434}{10^{-pH} + 10^{pH-14}} \tag{8-1}$$

where:

ΔN = difference in normality between the acid and base

ΔpH = change in pH

A common point of confusion is the combined use of a dead band or notch gain factor and a controller proportional band setting. In order for the user to understand the effect of these two settings, he should convert the proportional band to gain by inverting and multiplying by 100%. The gain of the controller's proportional mode is then equal to this gain outside the breakpoints and equal to this gain multiplied by the notch gain factor inside the breakpoints as shown by Eqs. (8-2a) and (8-2b). While the reset and rate modes are the same outside and inside the breakpoints, the contribution of these modes is multiplied by the proportional mode gain so that inside the breakpoints the contribution of all three modes to the controller output is reduced.

The effect of the proportional mode on the reset mode contribution is seen in the term "repeats per minute," which means repeats of the proportional mode contribution per minute.

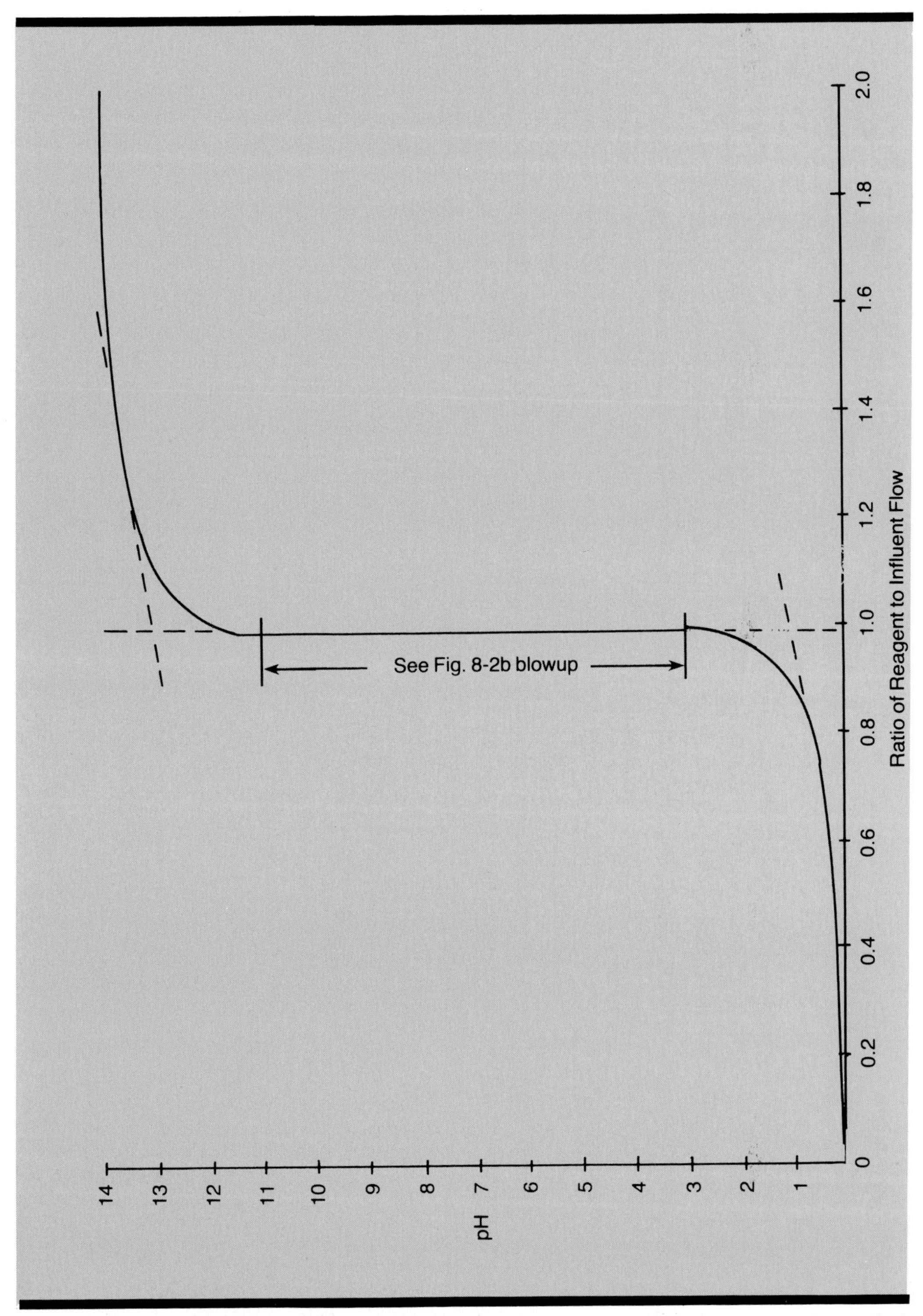

Fig. 8-2a. The straight line approximation for a nonlinear or notch gain controller may look good on a plot of the overall titration curve.

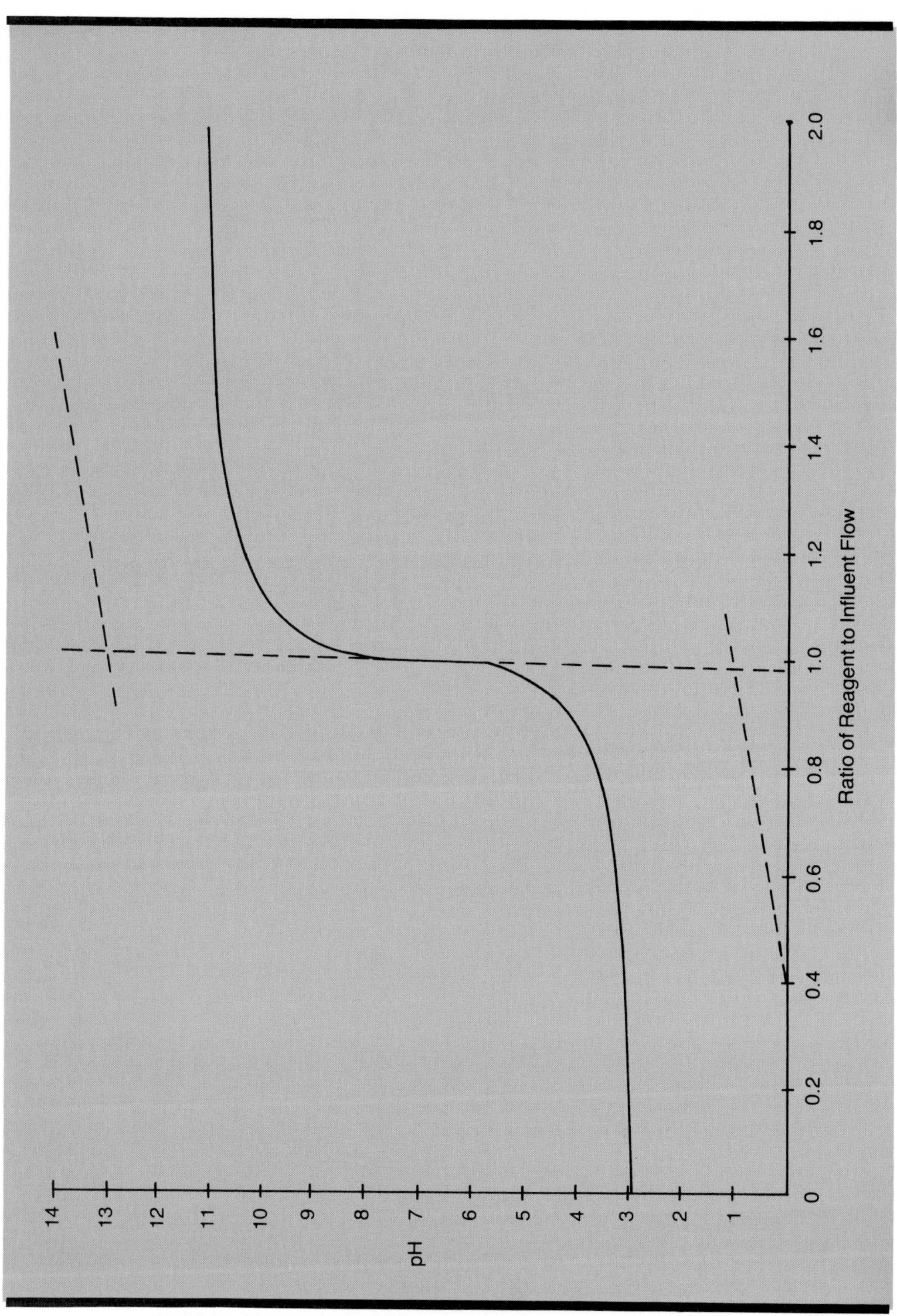

Fig. 8-2b. A blow up of the vertical region for the notch gain reveals another S-shaped curve and the misfit of the straight line approximation.

Outside the breakpoints:

$$K_c = \frac{100\%}{PB} \tag{8-2a}$$

Inside the breakpoints:

$$K_c = \frac{100\%}{PB} * K_n \tag{8-2b}$$

where:
K_c = controller's proportional mode gain

K_n = notch gain factor

PB = controller's proportional band setting (%)

Some microprocessor controllers have what the manufacturer calls an "adaptive" control algorithm that is actually a nonlinear controller with more than two breakpoints. The gain can be independently specified in three or more regions. Also, the regions as defined by the breakpoints can be based on a calculated value instead of a single measurement signal. The algorithm is called "adaptive" because the calculated value, which the breakpoints are based on, can be a function of operating conditions. However, the breakpoints and the associated gains are fixed. Since pH control uses pH for the breakpoints, nothing is free to be a function of operating conditions and the designation "adaptive" is a misnomer.

The minimum controller proportional band is proportional to the dead time to time constant ratio, the minimum integral time in minutes per repeat is proportional to the dead time, and the maximum derivative time is proportional to the dead time. (These relationships are described in Unit 9.) The equations in Unit 5 showed that the dead time and time constant for various types of mixed equipment depended on the volume and total input flow. If the equipment level and input flows are measured, the dead time and time constant can be calculated and the proportional, integral, and derivative modes can be corrected for equipment dynamics per the equations in Unit 9. These equations also show the effect of the steady state gains of loop components on the proportional mode-setting.

8-3. Signal Characterization

The more powerful microprocessor controllers have power functions and polynomials besides the math operations of addition, subtraction, multiplication, and division. Equations (2-18) through (2-25) can be programmed as function steps to construct a titration curve. The charge balance equation is solved for the reagent concentration that is manipulated by the pH controller. A direct solution of the reagent concentration in terms of pH is possible. The reagent concentration is converted from normality units to the units of the abscissa of the titration curve. The controller input signal is reagent demand, which compensates for the nonlinear gain from the titration curve. The controller output signal is multiplied by the influent flow to compensate for the nonlinear gain from the composition response. Unit 9 on controller tuning will show the cancellation of these gains in the open loop gain. The multiplication by the influent flow also provides a flow feedforward action where changes in influent flow result in immediate changes in valve signal without waiting for the pH feedback controller to see a pH error. If the calculated titration curve closely fits the actual titration curve, the pH controller is transformed into a linear reagent demand controller (Ref. 22). Figure 8-3 shows the arrangement of the required signal computations. Any horizontal shift of the titration curve is compensated for by reset action in the feedback controller. A vertical shift of the titration curve reduces the effectiveness of the linearization. The performance of a linear reagent controller for small vertical shifts or distortions of the curve is still much better than that of a conventional feedback controller unless the setpoint is right on a knee of the curve. If an online titration curve can be generated per Sec. 3-4, a process computer in the distributed control system network can be used to optimize the fit of the calculated titration curve. A standard Nelder-Mead subprogram can be used to search for the dissociation constants that minimize the error between the calculated curve and the identified curve in operating region. Each S-shaped curve requires the specification of at least one additional dissociation constant pH as shown in Fig. 8-4. The placement of an additional dissociation constant pH near an existing one will make a more gradual transition between the flat and steep portions of the curve. The fit of the calculated curve should be tailored to be most accurate in the control band.

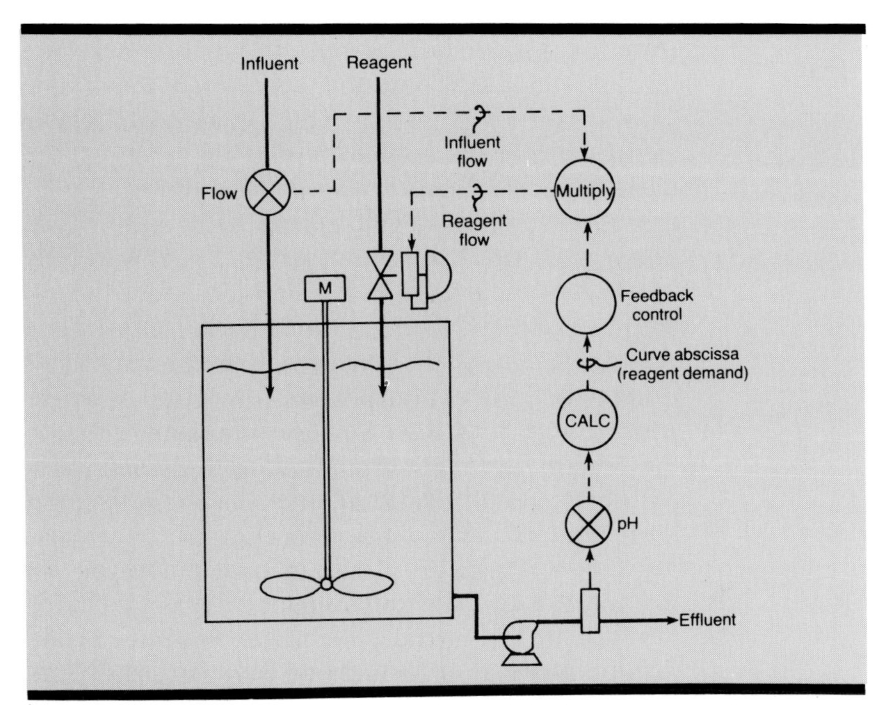

Fig. 8-3. The linear reagent demand controller calculates the abscissa of the titration curve from the pH measurement for use as the controller input and multiplies the controller output by the influent flow.

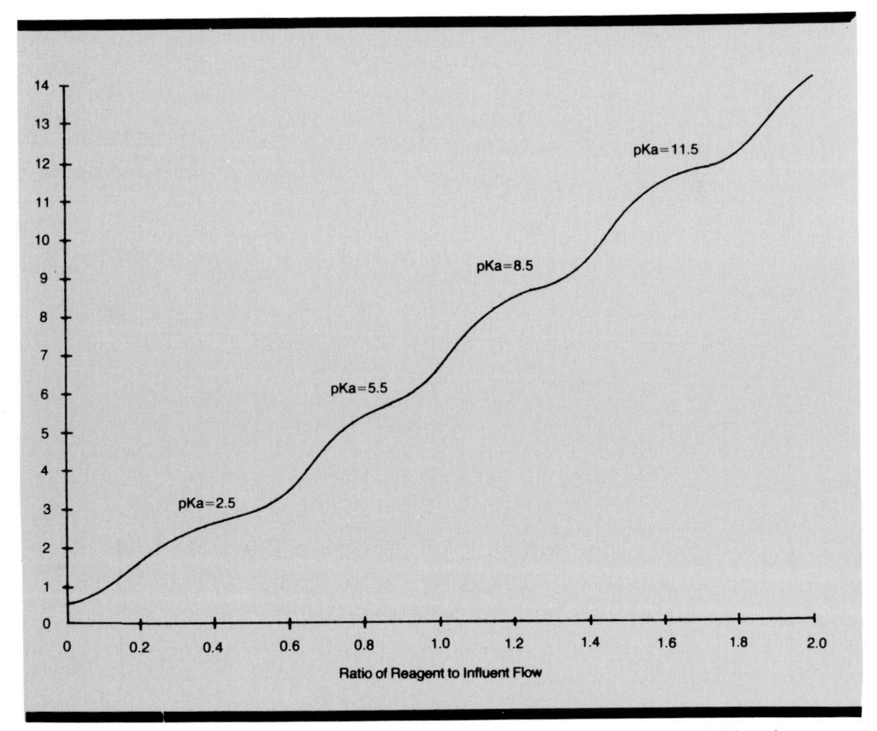

Fig. 8-4. Each S-shaped curve requires the specification of at least one additional dissociation constant pK to set the flat region.

A point of confusion in some of the pH control literature is the proper location of one steady-state gain to compensate for another steady-state gain. (The steady-state gain of a loop component is the output divided by the input after all transients have decayed). Specifically, some authors fall into the mistake of trying to compensate the nonlinearity of the titration curve by the use of a nonlinear valve gain. They improperly infer that a given nonlinear gain in the loop can be compensated for by another gain anywhere in the loop since this rule is true for linear gains. Since the pH transmitter span can be compensated for by a linear controller's proportional band or by a linear control valve's size, they infer that the decreasing slope as the pH measurement moves away from a setpoint at the equivalence point can be compensated for by the increasing slope of an equal percentage control valve characteristic. However, reset action in the controller breaks any fixed relationship between the controller input and output signals. Therefore, controller input nonlinear gains cannot in general be compensated for by controller output nonlinear gains and vice versa. The general procedure for steady- state gain compensation is as follows:

Steady-State Gain Compensation Procedure

(1) Plot the output signal of the component versus the input signal where the output is the ordinate and the input is the abscissa.

(2) If the plot is a straight line (constant slope), the gain is linear.

(3) If the plot is a curve (variable slope), the gain is nonlinear.

(4) If the gain is linear, the compensating gain can be located anywhere in the loop.

(5) If the gain is nonlinear and is located in the controller output or the process input, the compensating gain should be located in the controller output or in other words the controller output signal should be characterized.

(6) If the gain is nonlinear and is located in the controller input or the process output, the compensating gain should be located in the controller input or, in other words, the controller input signal should be characterized.

(7) The compensating gain is implemented with an equation to fit the plot on which the independent variable is the abscissa and the dependent variable is the ordinate. If the equation is exact, the product of the original gain and compensating gain is one.

The functional block diagram for an influent with two dissociation constants and a strong reagent that is suitable for programming in a microprocessor controller is shown in Fig. 8-5a. Each block corresponds to a functional step of the program. If the microprocessor does not have power functions but has polynomials, the third order polynomial approximation shown in Fig. 8-5b can be used up to 10 pH. Higher order polynomial approximations from the same source can be used if the operating region extends above 10 pH. About half of the steps in the program are for the polynomial approximation. Not shown but normally included in the program is the median selection from three pH measurements. A failure of one set of electrodes in any mode (fixed, upscale, or downscale reading) or the failure of two sets in opposite directions will not prevent the pH control system from functioning. Median instead of high or low signal selection is used because of the many failure modes of electrodes described in Unit 4-2.

For a strong acid and strong base system, it is not necessary to use dissociation constants in the signal characterization because the acid and base are completely ionized. The resulting calculation for the excess acid or base concentration is quite simple, as shown by Eq. (8-3) (Ref. 3). Single precision arithmetic will create a pH dead band around the equivalence point due to roundoff. However, the pH dead band due to control valve dead band is usually larger.

$$N_e = 10^{-pH} - 10^{pH-14} \qquad (8\text{-}3)$$

where:
N_e = excess strong acid or strong base concentration (normality)

pH = pH

The pH measurement can be trend recorded for monitoring. If the concept of a linear reagent demand is too confusing for operators so that the operators must see pH instead of reagent demand as the feedback controller measurement and setpoint, the engineer can program his own three-mode control algorithm to use

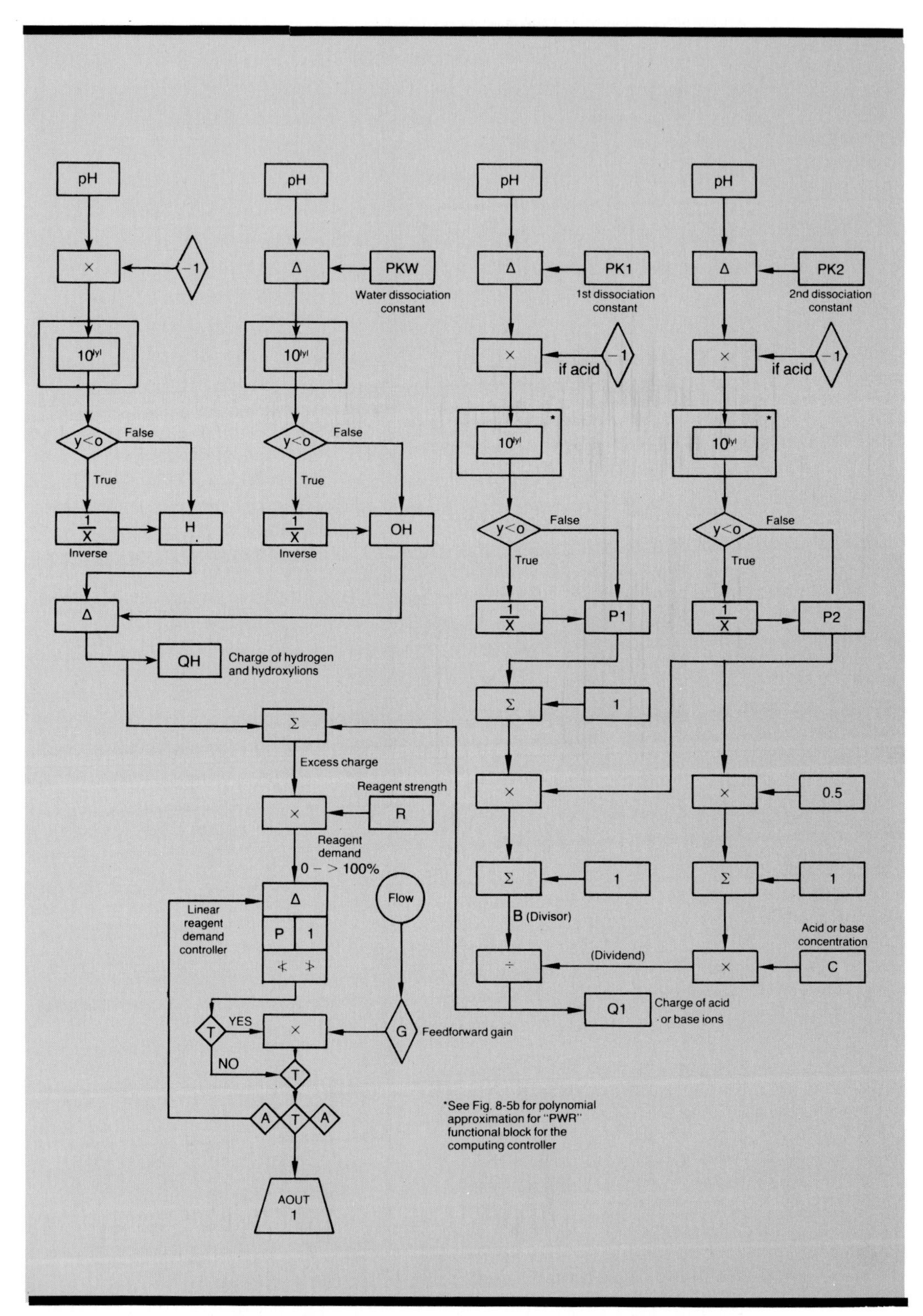

Fig. 8-5a. The titration curve can be programmed in a microprocessor controller by the use of functional step for each block.

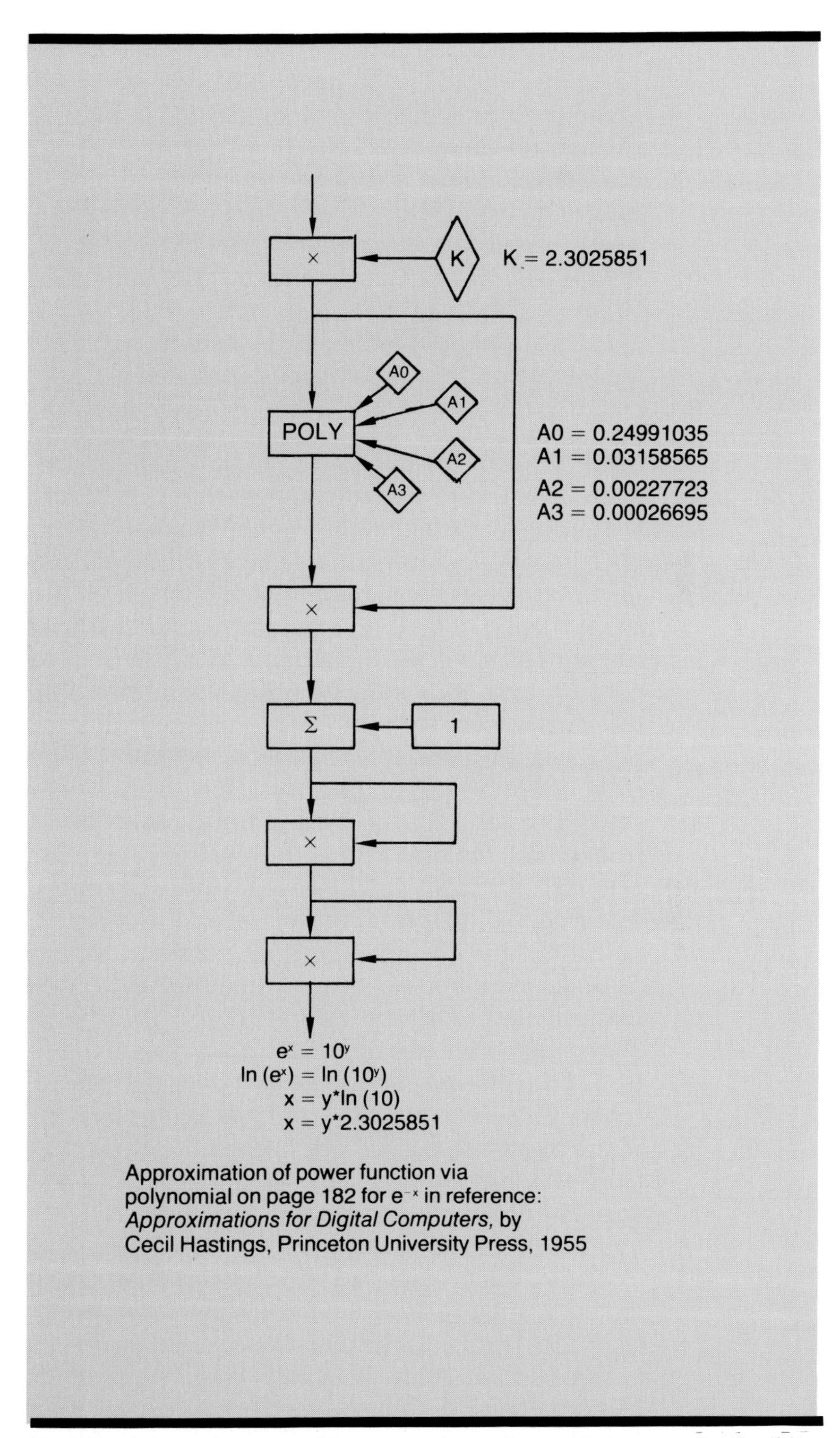

Fig. 8-5b. A third order polynomial approximation can be used for the power function if the power function step is not available.

reagent demand in the algorithm but display pH on the controller faceplate. The pH setpoint can be converted to the reagent demand setpoint by a polynomial fitted to the titration curve in the control band. However, the polynomial curve may reverse directions or blow up outside the fitted region. Limits should be placed on the pH input that coincide with the limits of the control band to protect against an accidental misplacement of the pH setpoint and consequential erroneous reagent demand setpoint. If the location or size of the control band is not fixed, the pH setpoint should be passed through the same math steps as the pH measurement to calculate reagent demand.

Leeds & Northrup developed the first microprocessor pH controllers that were preprogrammed to accept titration curve information for measurement signal characterization. While some pH controller manufacturers may claim signal characterization in verbal discussions, the actual method employed is usually some version of mode characterization with a fixed-gain relationship between the breakpoints. The Leeds & Northrup pH controllers calculate the reagent demand (abscissa of the titration curve) from the pH to obtain the controller input and multiply the controller output by the influent flow to obtain the reagent flow required. The Leeds & Northrup 7081 series pH controller has a family of titration curves, as shown in Fig. 8-6, from which the user selects the one that most closely fits the actual titration curve. The user specifies the upper portion "A" of the curve, the lower portion "B" of the curve, the equivalence point "C," the lower pH limit "D," and the upper pH limit "E." The user must insure he has accurately set the "D" and "E" limits to duplicate the titration curve in the control band since the curve shape changes for a blow up as shown in Figs. 8-2a and 8-2b. The titration curve emulated does not have to be symmetrical about the equivalence point, but it must have a single S-shape in the operating region. The Leeds & Northrup 7084 series pH controller has the ability to switch between two titration curves based upon the status of remote contacts which enables it to handle a drastic change in influent or reagent composition. For example, waste ammonia and its titration curve would be selected over purchased caustic and its titration curve as a reagent until the supply of ammonia was depleted. The titration curves are based upon linear interpolation between 16 data points entered by the user. The 7084 series can also compensate for the change in dissociation constants, and hence effluent pH with temperature, and can characterize the output signal to compensate for the nonlinearity of an equal percentage control valve. Since data points are entered, the titration curve can have any shape desired.

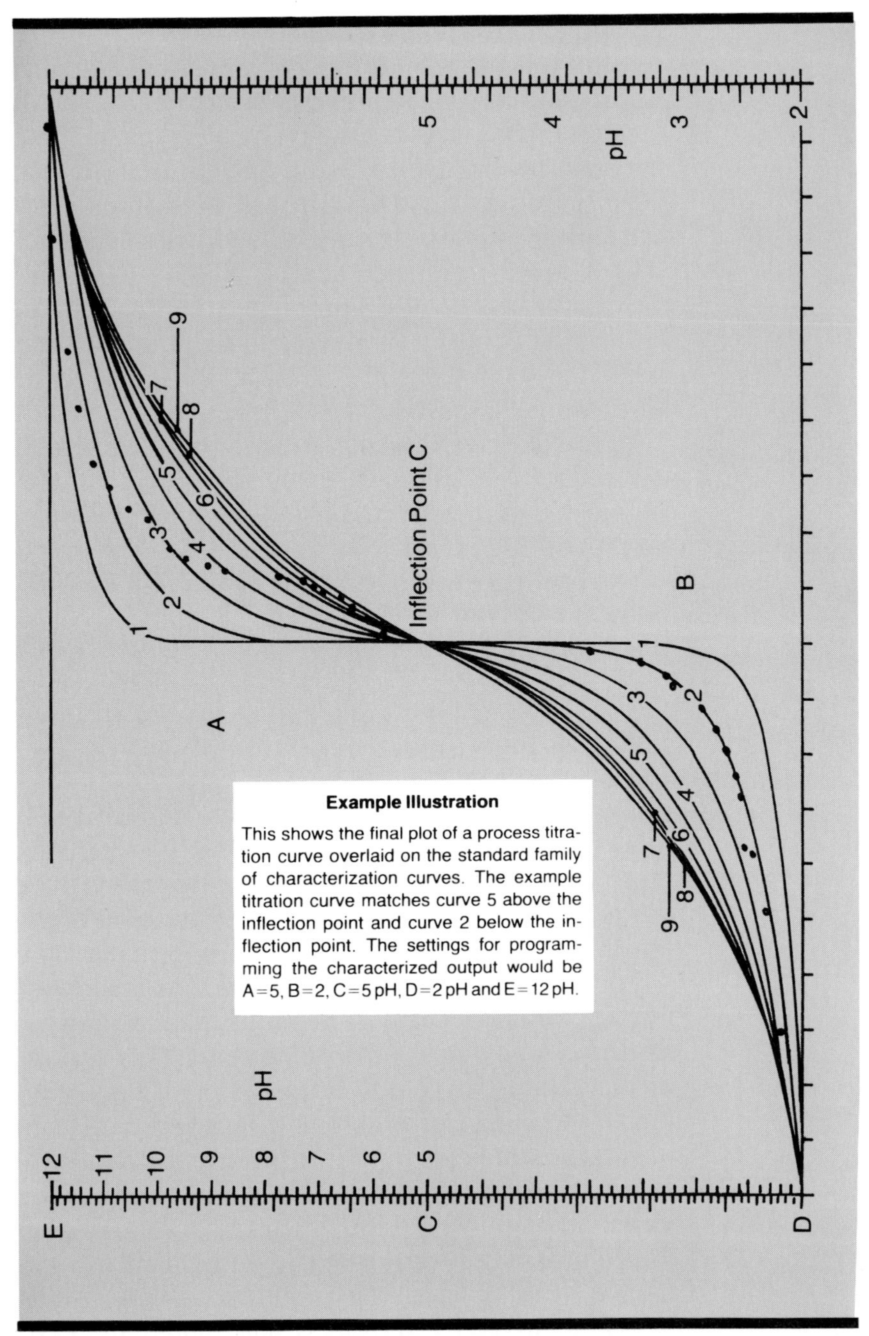

Fig. 8-6. The Leeds & Northrup 7081 series pH controller has a series of preprogrammed titration curves from which the user can select one that matches his actual curve for signal characterization.

Signal characterization facilitates the full attainment of the potential of feedback control allowed by the dead time to time constant of the control loop. For a strong acid and strong base system, the controller proportional band cannot be set high enough and the notch gain low enough to prevent continuous oscillations between 2 and 12 pH without signal characterization. Also, the time required to reach the first peak after startup is about 13 times larger without signal characterization (Ref. 4). It is very difficult to manually obtain an initial pH in a vessel that is near the control band prior to startup of the control system. If it is difficult to titrate a strong acid and strong base in the laboratory with a burrett, you can imagine what it's like with a control valve. The same startup problem exists to a lesser degree in pH systems with buffering if the setpoint is on the steep portion and the pH measurement is on the flat portion of the titration curve. The change in the distance of the setpoint from the measurement in pH units seen by a conventional pH feedback controller does not reflect the change in distance in reagent units. The result is sluggish action when the pH measurement is far out on the flat portion of the titration curve whether due to startup or a large disturbance if characterization is omitted. Figures 8-7a and 8-7b show the greater stability and faster startup obtained by signal characterization.

8-4. Feedforward Control

Flow feedforward is straightforward and relatively easy to implement unless the influent flow is difficult to measure. For example, there is no easy way to measure flow in a partially filled underground sewer line. Short-term influent flow measurement errors or noise should not cause a reagent error greater than the allowable reagent error band. Long-term feedforward flow signal errors should be corrected by multiplication instead of addition by the pH feedback controller output. This allows the pH controller to stop reagent flow when the influent stream has no acid or base and to stop reagent addition for no influent flow by multiplication by zero. Flow feedforward can be viewed as flow ratio control where the ratio is corrected by the output of the pH feedback controller. Flow feedforward is also a part of the linear reagent demand controller shown in Fig. 8-3.

An influent flow disturbance will cause a peak error for closed loop inline pH control that is about equal to the open loop error because of the small disturbance and equipment time constants.

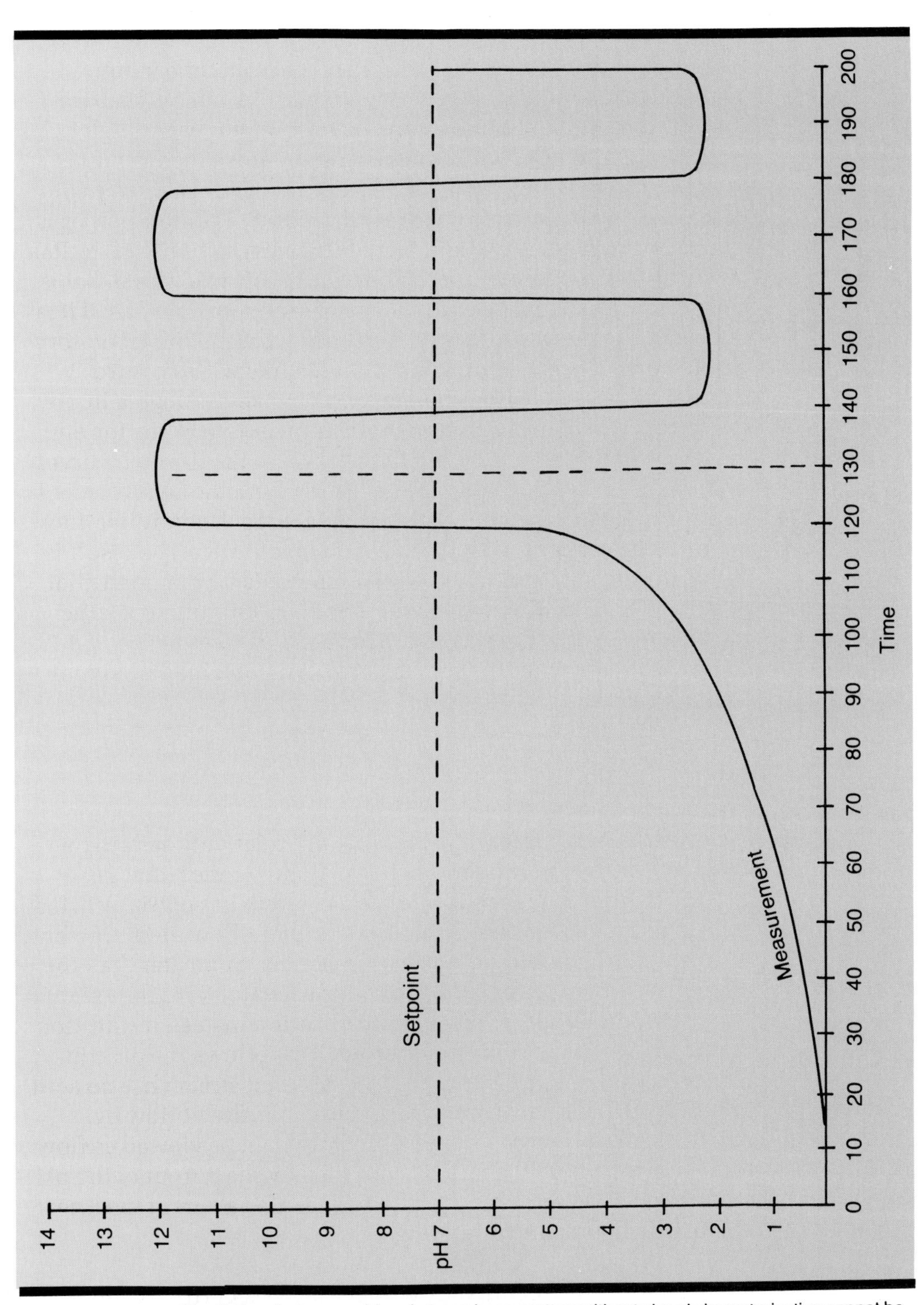

Fig. 8-7a. A strong acid and strong base system without signal characterization cannot be stable between 2 and 12 pH and will take a long time to startup or recover from a large disturbance.

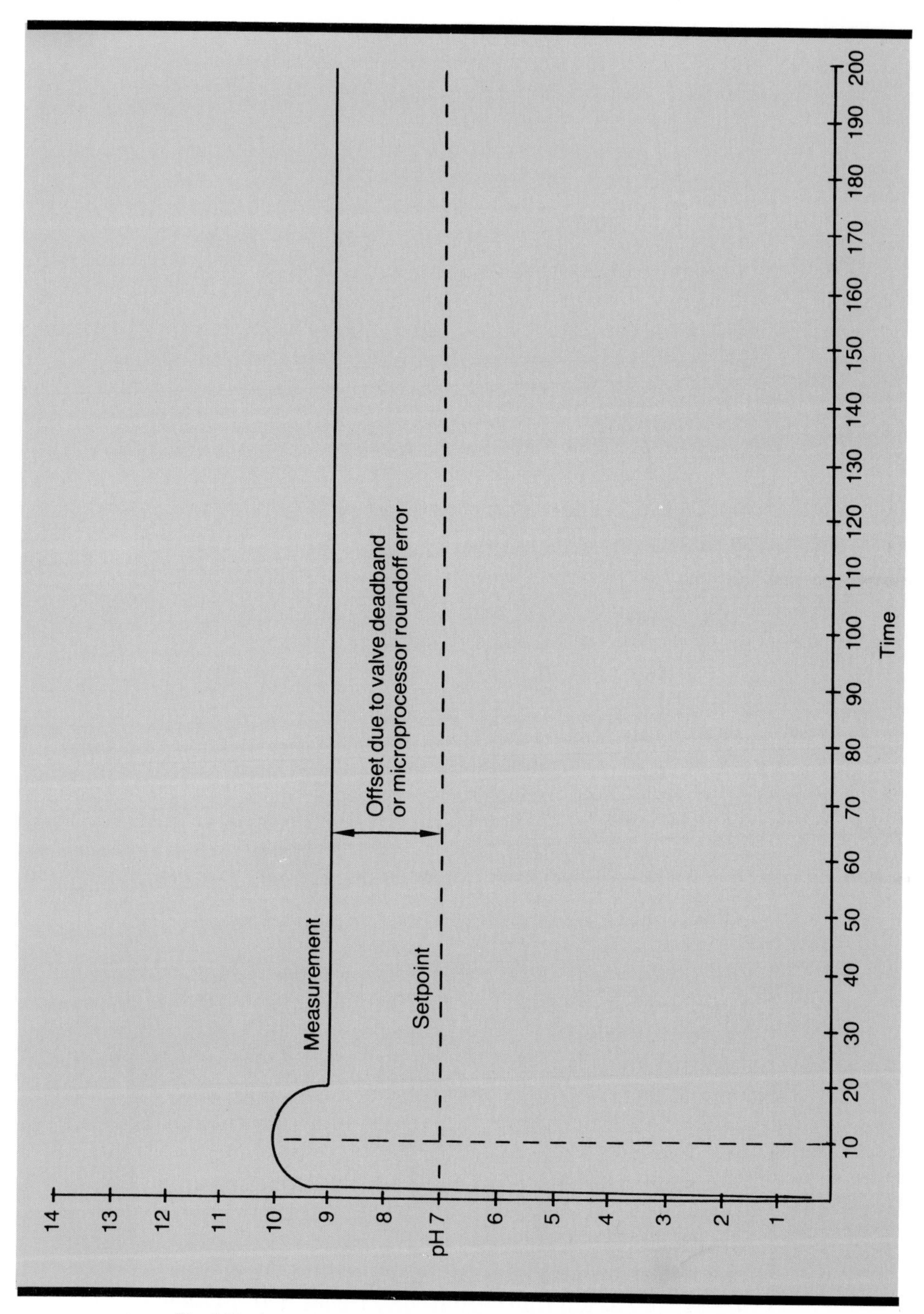

Fig. 8-7b. A strong acid and strong base system with signal characterization can be stable between 2 and 12 pH and will take a short time to startup or recover from a large disturbance.

The resulting peak error for the flow disturbance will only last for a few seconds and may be adequately attenuated by the filtering effect of downstream volumes. If the attenuation downstream is not great enough, flow feedforward or ratio control, as shown in Fig. 8-8 should be used. The influent flow measurement should be close enough to the static mixed to prevent the flow correction signal from arriving sooner than the actual flow upset. The inline pH feedback loop will counteract the feedforward correction if the actual influent flow increase at the mixer is delayed by more than the dead time of the inline pH loop (about one to four seconds). Fortunately, the influent flow pipeline runs full so that the liquid flow response dead time is less than a fraction of a second. The usual cause of an excessive flow response dead time is an intervening piece of equipment between the influent flow measurement and the inline pH system. While a dead time or lag time could be added to the feedforward signal for this situation, it is difficult to set the time accurate enough to do much good.

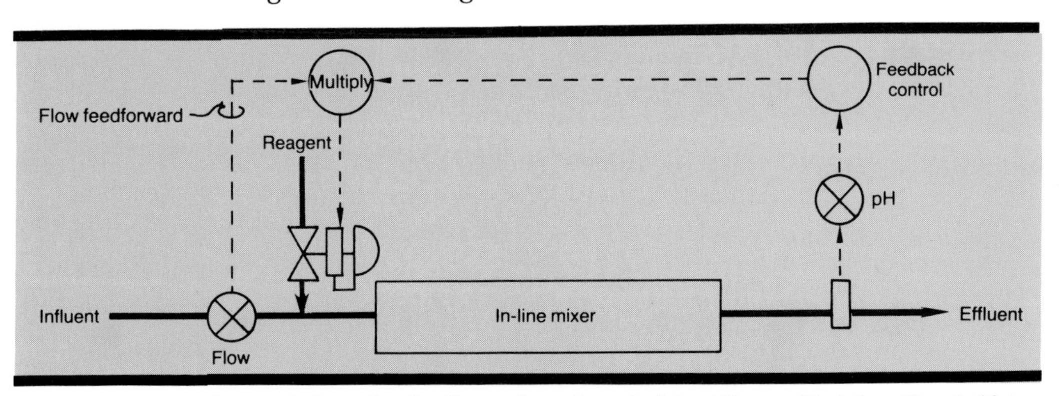

Fig. 8-8. Influent flow feedforward or ratio control should be used for inline pH control for fast flow disturbances that are not attenuated enough by downstream volumes.

pH feedforward control was difficult to implement and was apt to do more harm than good before the advent of the microprocessor controller. Various attempts were made to provide a feedforward ion concentration signal with analog modules or equal percentage control valves that was difficult to modify and keep calibrated. The computational power, accuracy, and flexibility of the microprocessor facilitated the development of a method that suffers less from performance degradation and is conceptually simpler and more effective. This method is based on the titration curve and is identical to the method used for signal characterization. In fact, the feedforward signal generator can be visualized as a proportional-only linear reagent demand controller with a proportional mode gain equal to one (Ref. 5). The reagent flow per influent flow required to go from the influent pH to the setpoint pH is calculated as the abscissa of the titration curve programmed into the microprocessor. This signal is then multiplied by the influent flow to provide a signal that is reagent flow demand to reach setpoint. This reagent flow demand can be corrected by the addition of the deviation of a pH feedback controller's output from 50%, as shown in Fig. 8-9a. The feedback controller is able to make plus and minus corrections to the reagent flow demand. If the distance in reagent units from the influent point to the setpoint is large and the allowable reagent error band is small, valve position control can be used to correct the feedforward reagent demand signal to a large control valve and the output of the pH feedback controller can be used to position a small trim control valve, as shown in Fig. 8-9b. If the feedforward signal is accurate, the pH will always be near the control band and a conventional nonlinear or notch gain pH controller can be used for feedback trim. The accuracy of the titration curve is always in question for waste treatment applications. The accuracy of the influent pH measurement is a concern for high and low pH due to alkaline error and acid error but also due to the flatness of the titration curve at its extremeties. A small pH measurement error translates to a large reagent demand error on the flat portions of the titration curve, as shown by Fig. 8-10. Alkalinity or acid error will cause the feedforward reagent demand to be much less than required since the measured pH will read lower and higher, respectively, than the actual pH. Therefore, pH feedback control should always be used to correct the pH feedforward signal.

Leeds & Northrup uses a proportional-only 7084 series controller and influent pH measurement to generate a feedforward signal and a three mode 7084 series controller and effluent pH measurement to generate a feedback signal. The two signals are combined by an add/multiply algorithm to position the reagent control valve. Leeds & Northrup calls this scheme "adaptive control" because the feedforward signal is adapted by the feedback signal (Ref. 6).

8-5. Cascade Control

The use of an inner reagent flow control loop can help correct for reagent flow disturbances caused by pressure disturbances and control valve hysteresis and fouling. However, an upstream pressure regulator could be used to reduce pressure disturbances, a positioner could be used to reduce hysteresis, and the pH controller can correct for fouling because it develops slowly. These alternatives should be used wherever possible because there are several potential problems in the application of pH to flow cascade control.

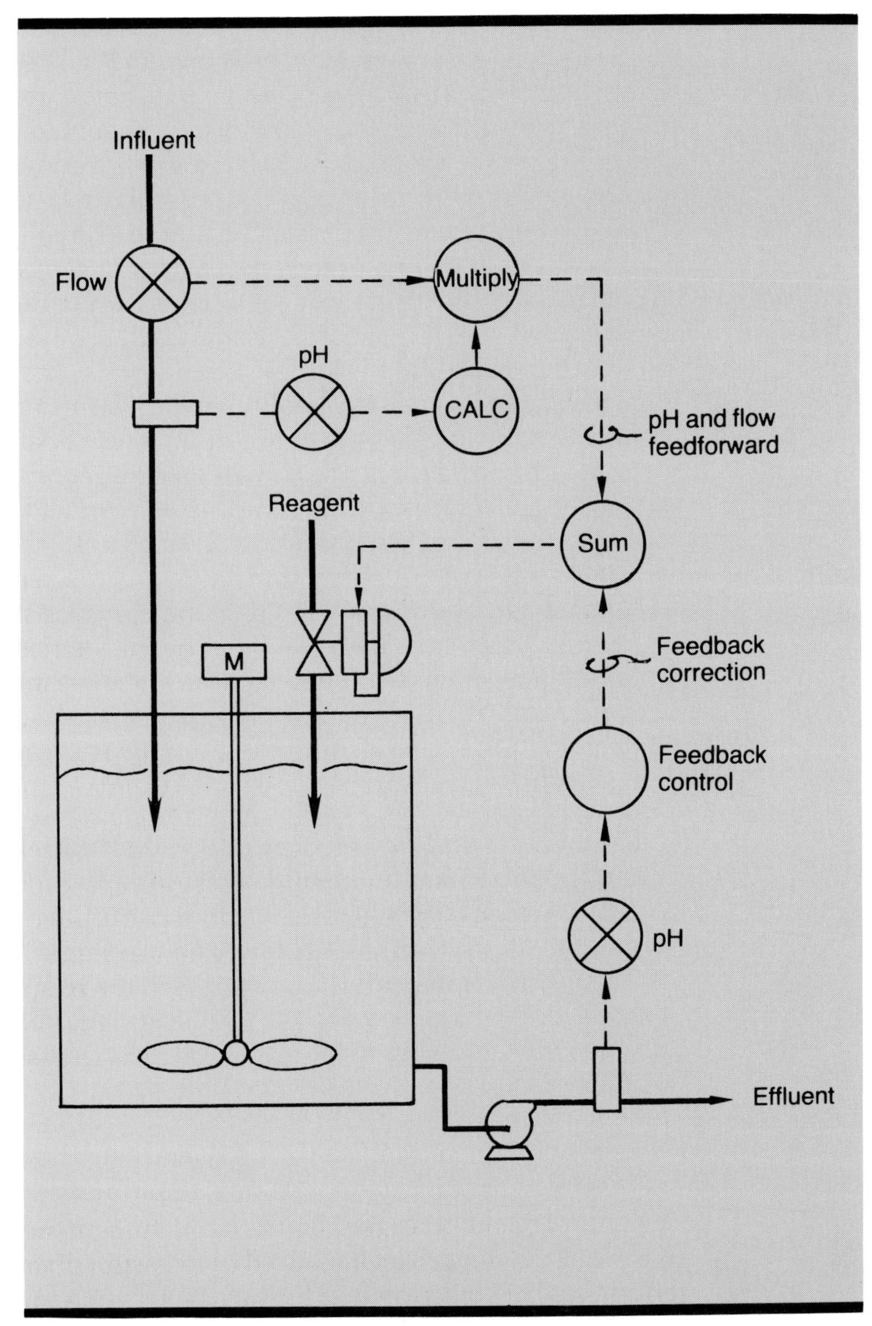

Fig. 8-9a. The pH feedback controller can correct the feedforward signal to a single large control valve if the large valve's hysteresis is not too large.

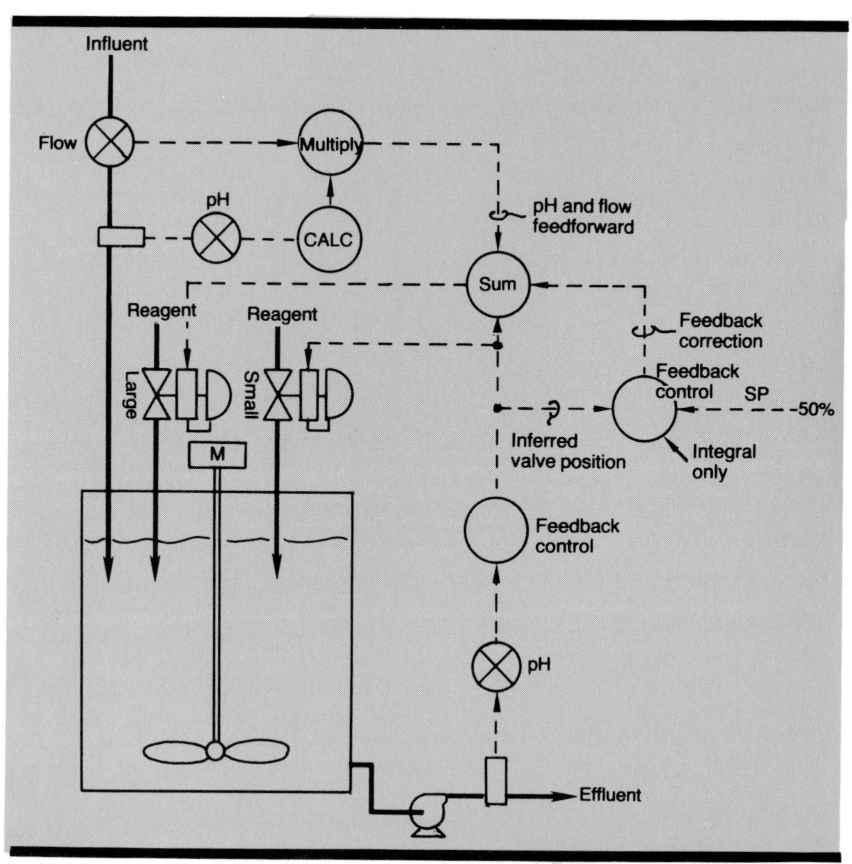

Fig. 8-9b. The pH feedback controller can position a small trim valve and valve position control can correct the feedforward signal to the large valve if the large valve's hysteresis is too large.

For pH to flow cascade control, the reagent flow measurement noise and repeatability requirement is similar to the control valve hysteresis requirement and the flow measurement rangeability requirement is identical to the control valve rangeability requirement. The flow controller will cause control valve dither if the proportional band is not increased and will cause a slower reagent delivery response if the proportional band is increased for flow measurement noise. Equation (6-1c) can be used to estimate the allowable error band due to measurement noise or repeatability and Eq. (6-3b) can be used to estimate the required flow measurement rangeability. Orifice plates have only a 4:1 rangeability and may have a 5% noise band and are therefore not recommended for cascade control in pH systems. Most flow measurements have a rangeability less than the rangeability of a globe control valve with teflon packing, a high performance positioner, and a constant upstream pressure. Table 8-1 summarizes the approximate rangeability of some meters for low

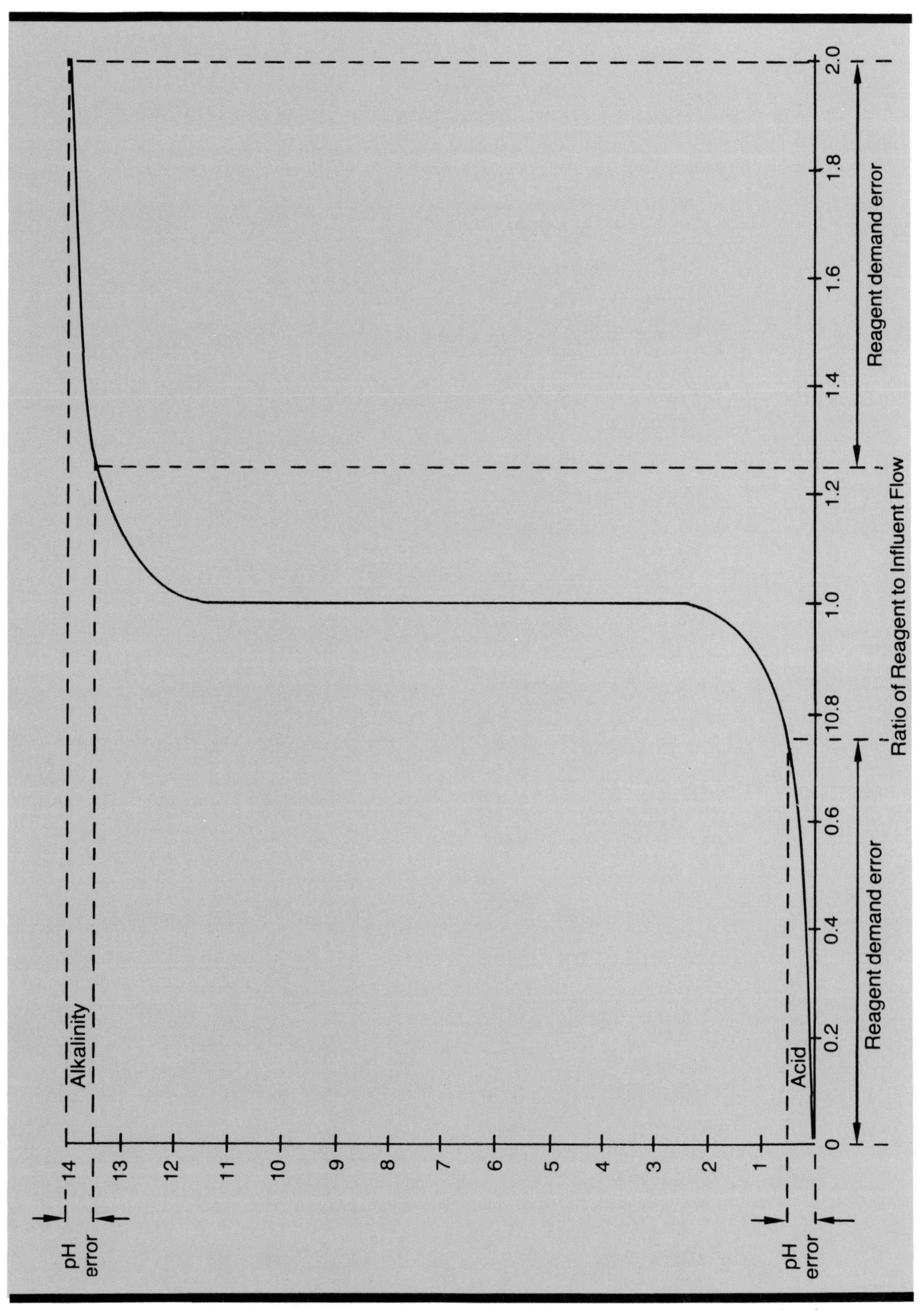

Fig. 8-10. The akalinity and acid pH measurement errors translate to large feedforward signal errors due to the flatness of the titration curve.

flow measurement. The rangeability of the magmeter and vortex meter listed in the table is only achievable for a maximum reagent flow requirement that happens to equal the meter's maximum flow for a given line size (such a coincidence is rare).

Type of Flowmeter	Rangeability
Differential head (orifice)	4:1
Rotometer	11:1
Vortex meter	15:1
Turbine meter	10:1
Turbine meter with extended range electronics	40:1
Magmeter	50:1
Gyroscopic mass flow meter	100:1
Helix positive displacement	150:1
Piston positive displacement	400:1

Table 8-1. The rangeability of many conventional meters for low flow measurement is insufficient for pH to flow cascade control.

The use of a positioner on a reagent flow loop creates an inner valve position feedback loop that is slower than the outer loop. The use of pH to flow cascade on an inline pH system creates an inner loop that is slower than the outer loop. The outer controller of both of these cascade loops must be detuned to prevent persistent oscillation (Ref. 8).

While a pH to flow cascade loop is not recommended for most applications, the installation of a reagent flow meter is extremely helpful in trouble shooting the reagent delivery system. A reagent flow measurement can help identify the following problems:

Reagent Valve Problems Identified by a Flow Measurement

(1) Plugging of the control valve due to small trim size.

(2) Leaking of the control valve due to corrosion or wear from pulsing.

(3) Erratic flow characteristic of control valve due to hand-cut trim.

Instead of doing pH control on a poorly mixed vessel, an inline pH system can be installed upstream and the poorly mixed vessel used as a filter to attenuate the fast oscillations of the inline system. The result is a performance level in terms of peak error that is as good as a feedback loop on a well-mixed tank with a residence time about equal to the equipment time constant of the poorly mixed tank and an equipment dead time about equal to the inline loop dead time. (Unit 10 will compare the performances of various vessel types.) However, there is no guarantee that the attenuated oscillations will average out to the pH setpoint even if they are centered about the inline setpoint due to the nonlinearity from the titration curve. A pH to pH cascade control system is sometimes used to correct this problem. A feedback pH loop on the outlet of the poorly mixed vessel slowly trims through integral mode action the setpoint of the inline pH loop, as shown in Fig. 8-11. The inline setpoint correction may take several hours for a sump and several days for a pond or lagoon. Therefore, the inline system must be located to handle all of the disturbances so that the sole purpose of the outer pH loop is correction for the long-term, nonlinear effects of the titration curve. A digital electronic filter cannot be used to emulate the vessel filter because the electronic filter attenuates the pH measurement oscillations whereas the vessel attenuates the concentration oscillations. The nonlinearity of the titration curve assures their nonequality.

8-6. Adaptive Control

As previously mentioned, some manufacturers call their nonlinear or notch gain controllers or combination feedforward and feedback controllers "adaptive" controllers. In this text, a controller will be called adaptive only if it identifies and compensates for the dynamics and steady-state gains of the pH process.

All pH systems with a constant titration curve and reasonable reagent dissolution time can be controlled if the proper number and type of mixing equipment and controls are installed and if the pH measurement can be accurately and reliably made. Adaptive control can improve the performance of loops with a variable titration curve. The greatest benefit is achieved when the slope in the operating region of the titration curve suddenly becomes steep. While adaptive control can also help identify and compensate for a long reagent dissolution time, there is still a conversion problem, per Fig. 7-1, to be reckoned with.

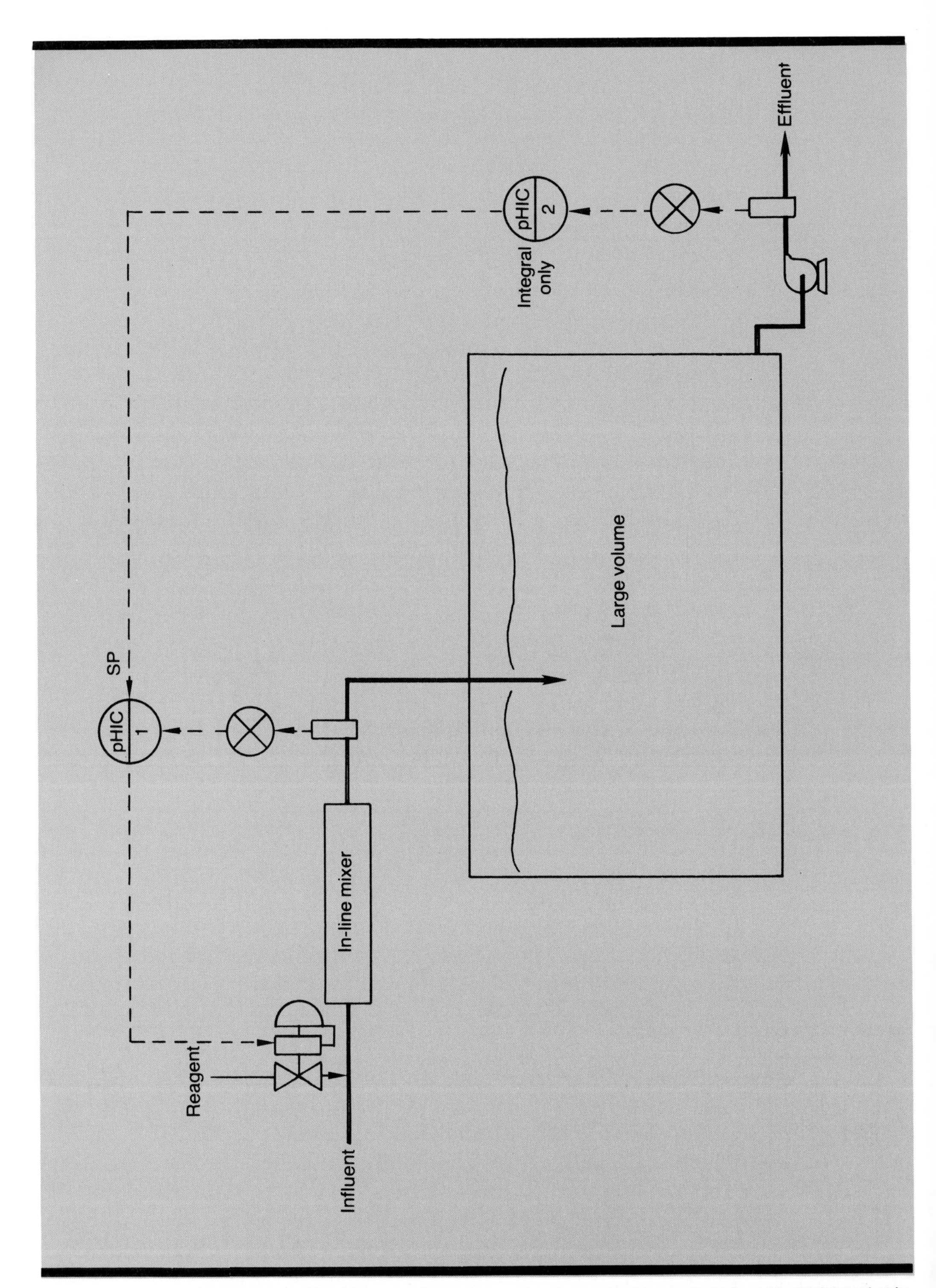

Fig. 8-11. A pH to pH cascade control system is used to trim the inline loop setpoint to slowly compensate for pH nonlinearity that causes an offset of the attenuated effluent pH from setpoint.

The Foxboro Nonlinear/Adaptive Control Package is the only commercially available pH controller to date that is adaptive in the sense defined in this text. It consists of a nonlinear controller and an adaptive control unit, as shown in Fig. 8-12. The adaptive control unit monitors pH loop oscillations and distinguishes between high and low frequency oscillations in a discriminator unit. High-frequency oscillations, such as limit cycling, cause a positive DC correction signal proportional to the amplitude and frequency of the oscillations. This signal is integrated and used to widen the dead band width of the nonlinear controller. Low frequency oscillations, such as a slow drift from setpoint, cause a negative DC correction signal that is used in a similar fashion to narrow the dead band width. The adaptive control unit has three settings and the nonlinear controller has four settings to adjust. The cross-over frequency setting of the adaptive unit should be set at approximately three times the ultimate period of the loop. The integral time setting of the adaptive unit should be set approximately equal to the integral time of the nonlinear controller. The low frequency gain setting of the adaptive unit should be set high enough to prevent drift but low enough to prevent burst cycling (Ref. 7). The nonlinear controller dead band gain setting is determined from the titration curve as described in the unit on mode characterization. The proportional, integral, and derivative mode settings are determined by techniques described in the next unit for tuning conventional feedback controllers for pH processes. The main advantages of this package are the industry application experience and technical support by Foxboro. The main disadvantages are the large number of settings and the limited compensating power of the dead band width adjustment. The greatest improvement is realized when the slope of the titration curve in the operating region changes by the same ratio as the proportional mode to dead band gain ratio.

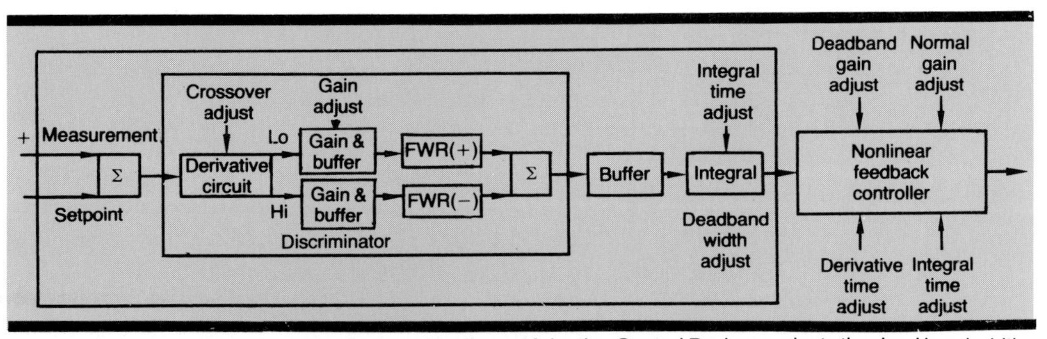

Fig. 8-12. The Foxboro Nonlinear/Adaptive Control Package adapts the dead band width to a changing titration curve.

Nearly all of the adaptive control schemes described in the literature attempt to identify a single pH process gain rather than the continuous change in pH process gain represented by the titration curve. While this single process gain, when properly identified, is sufficient to insure loop stability, it continually changes with pH, even if the curve is fixed. The pH process gain seen by the pH controller is the difference in pH divided by the difference in reagent demand per influent flow between the measurement and setpoint. This is represented by the slope of a line drawn from the setpoint to the measurement (Ref. 8). Figure 8-13 shows how the slope of this line, and hence the gain, varies as the pH measurement moves along the titration curve. An adaptive control strategy that attempts to identify this gain or a similar gain is forced to continually readjust the gain as the pH changes, even though the shape of the titration curve hasn't changed. The adaptive control system is constantly trying to play catch up.

An adaptive control system can only identify what it sees. The larger the pH oscillations, the greater the region of identification. To avoid upsetting the effluent pH, many adaptive schemes use a small auxiliary vessel or inline system with a pH control system that is perturbed to create pH oscillations for identification. The auxiliary influent flow is small compared to the main influent flow so that auxiliary pH oscillations don't upset the effluent pH when the streams are recombined. Auxiliary inline systems provide faster identification, but the accuracy of the identification suffers from measurement noise and nonlinear valve and electrode dynamics. The reader is directed to Ref. 10 for a summary of various adaptive control schemes documented in the literature.

If the controller input is reagent demand instead of pH, the control action accelerates as the pH moves to a flat portion of the titration curve when compared to the control action of a linear or nonlinear pH controller. Signal characterization provides a faster startup and recovery from large disturbances. Also, the process gain to be identified does not change as the pH moves along the curve for a fixed curve. Thus, a better adaptive control scheme would seek either to identify the change in reagent demand gain or the actual change in the titration curve shape for the linear reagent demand controller described in the unit on signal characterization. An easily envisioned method would periodically identify the titration curve by the pipeline method shown in Fig. 3-8, and subsequently update the curve by using the new data points and linear interpolation or by using a search

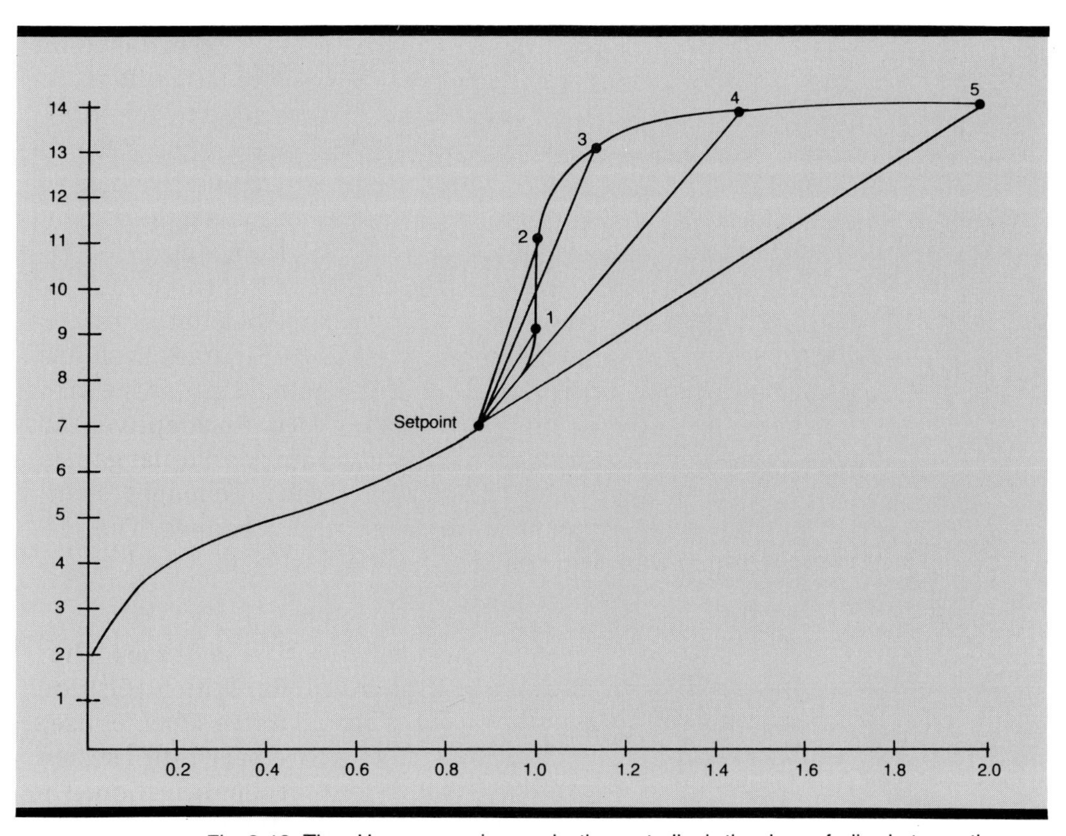

Fig. 8-13. The pH process gain seen by the controller is the slope of a line between the setpoint and the pH measurement on the titration curve.

technique to update the dissociation constants to optimize the fit of a calculated curve to an identified curve.

A Smith Predictor might be effectively employed to help suppress control loop oscillations for an inline pH system with a fast measurement and reagent delivery system. If the measurement and reagent systems are fast enough so that changes in their dynamics are negligible compared to the transportation delay of the main stream, the loop dead time is well defined and is inversely proportional to the throughput flow. A Smith Predictor will provide better performance and suppress oscillations for moderate gain changes as long as the dead time estimate is accurate. The loop performance is better because the inline controller cannot use any derivative action and the loop dead time to time constant ratio is relatively large (3:1 or more). Figure 8-14 shows the addition of a Smith Predictor to a pH controller output with the dead time calculated as the equipment volume divided by the influent and reagent flow and biased by the total of the instrument dead times.

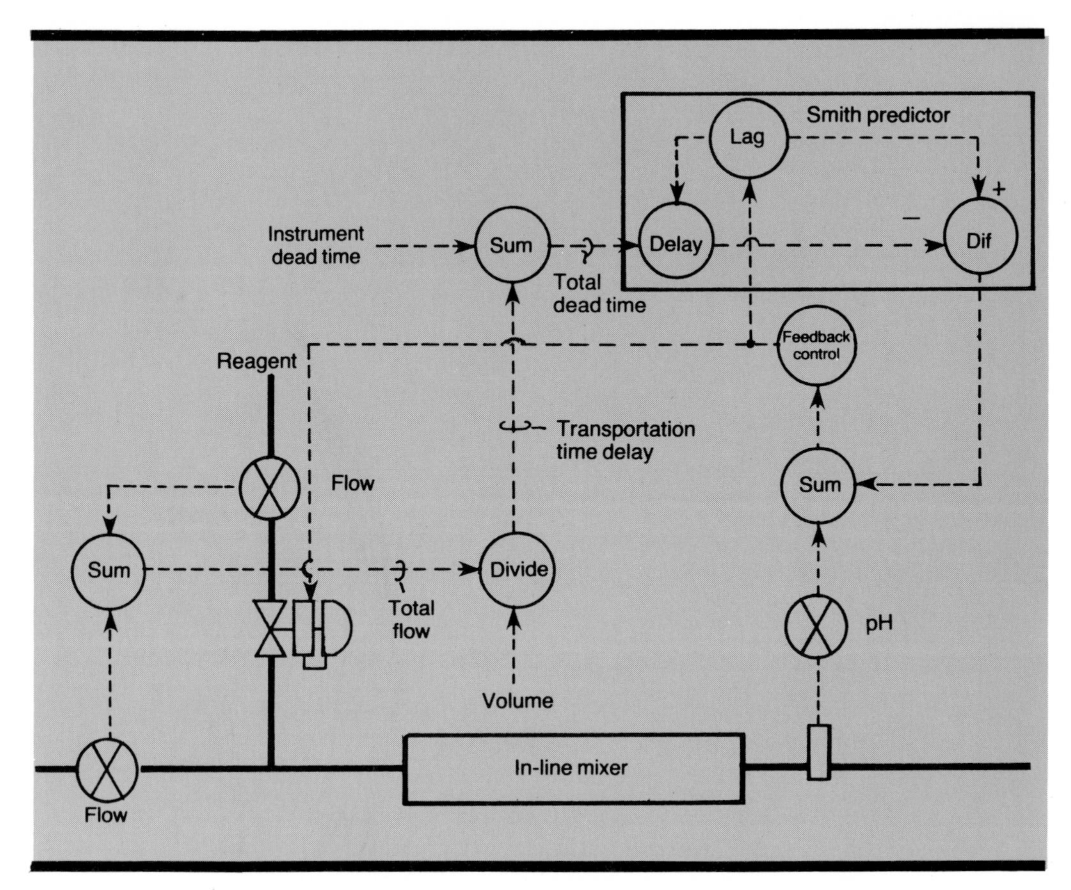

Fig. 8-14. A Smith Predictor can be added to an inline pH system to help suppress oscillations for moderate gain changes if the changes in the instrument dynamics are negligible.

Exercises

8-1. *What happens to the time to the first peak for a given disturbance to a pH feedback loop if the electrode becomes coated?*

8-2. *What advantages does a linear reagent demand controller have over a nonlinear controller?*

8-3. *Why would a pH feedforward controller be added?*

8-4. *If the influent pH is below 0 pH, should pH feedforward be used?*

8-5. *If the spray water to the top of an absorber needs to be pH controlled and the pH is measured at the discharge of the circulation pump, should the reagent be added to the spray as it enters the column, to the sump, or to the suction of the circulation pump?*

8-6. *Why should a linear reagent demand controller be used for adaptive control?*

References

[1]Shinskey, F. G., *pH and pION Control in Process and Waste Streams*, John Wiley & Sons, 1973, p. 56.
[2]Gray, D. M., "Microprocessor Characterizes pH ahead of Controller for Easy Tuning," *Control Engineering*, Jan. 1980, pp. 79-80.
[3]McMillan, G. K., "Improved Control by Signal Linearization," ISA Conference Proceedings, March 1981, Paper 521, p. 66.
[4]*Ibid*, pp. 67.
[5]Gray, D. M., "Characterized Feedback and Feedforward pH Control," ISA Transactions, Vol. 20, No. 2, 1981, p. 65-66.
[6]Leeds & Northrup 7084 Microprocessor pH Analyzer/Controller Operator's Manual, No. 277610, Rev A., pp. 3, 24.
[7]Foxboro Supporting Literature SI 1-00745, Spec 200 Nonlinear/Adaptive Controller Package for pH Control System, July 1973, pp. 1-10.
[8]Shinskey, F. G., *pH and pION Control in Process and Waste Streams*, John Wiley & Sons, 1973, pp. 140-141.
[9]McMillan, G. K., *Tuning and Control Loop Performance*, ISA Monograph 4, 1983, pp. 193-194.
[10]Trevathan, V. L., "Advanced Control of pH," *ISA Advances in Instrumentation*, Vol. 33, Part 2, 1978, pp. 72-73.

Unit 9: pH Controller Tuning

UNIT 9

pH Controller Tuning

This unit describes how the ultimate period and controller mode settings depend on the dynamics and steady-state gains of the loop components.

Learning Objectives—**When you have completed this unit you should:**

A. Know how to estimate the ultimate period from the loop dead times and time constants.

B. Know how to estimate the controller mode settings from the ultimate period, major time constant, and steady-state gains.

C. Understand the effect of the pH process gain nonlinearity on the mode settings.

9-1. Ultimate Period

The ultimate period or natural period of the loop is the period of the equal amplitude oscillation obtained by decreasing the proportional band with the integral and derivative modes turned off (no reset or rate action). If the oscillation amplitude is growing, the measured period is shorter and if the oscillation amplitude is decaying, the measured period is longer than the ultimate period. The ultimate period can be estimated by Eq. (9-1) for continuous pH control. Note that ultimate period is equal to about two times the loop dead time for large dead time to time constant ratios and is equal to about four times the loop dead time for small dead time to time constant ratios (Ref. 1).

$$T_u = 2 * \left[1 + \left[\frac{TC}{TC + TD} \right]^{0.65} \right] * TD \qquad (9\text{-}1)$$

where:

TC = largest time constant in the loop (minutes)

TD = loop dead time (minutes)

T_u = ultimate period for continuous control (minutes)

For batch pH control, there is no effluent flow; consequently, the composition response is integrating to a step change in reagent or influent composition or flow. There is no negative feedback, and hence no self-regulation. The composition response will ramp until a physical limit is reached but the movement of the operating point from the steep to the flat portion of the titration curve may give the false impression that the pH is "lining out" and the process is self-regulating. The ultimate period can be estimated by Eq. (9-2a) for batch pH control. Note that the ultimate period is equal to about four times the dead time for large dead time to time constant ratios. For a given dead time and time constant, the ultimate period will be significantly larger for the batch system. For example, if the dead time is 0.1 minute and the time constant is 1 minute, the ultimate period for continuous pH control is about 4 times the dead time or 0.4 minutes whereas the ultimate period for batch pH control is 22 times the dead time or 2.2 minutes (Ref. 2).

$$T_u = 2 * \left[1 + \left[\frac{TC}{TD} \right]^{0.65} \right] * TD \qquad (9\text{-}2)$$

where:

TC = largest time constant in the loop (minutes)

TD = loop dead time (minutes)

T_u = ultimate period (minutes)

A pH loop has many individual dead times and time constants, as shown in Fig. 8-1. The loop dead time is the total of all the pure time delays plus the equivalent dead time of all time constants in the loop smaller than the largest (Ref. 3). The largest time constant is the equipment time constant except for inline systems and poorly designed lime systems where the dissolution time exceeds the residence time. The equivalent dead time for each smaller time constant is found by multiplying the smaller time constant by the effective dead time factor from Fig. 9-1. Note that nearly all of the time constant is converted to dead time if the ratio of the small to large time constant is small. For vessels, sumps, ponds, and lagoons, the instrument time constants are so small relative to the equipment time constant that they can be summed directly as equivalent dead time. For an inline system, the largest time constant is associated with either the measurement electrode or control valve response. Both of these time constants are nonlinear and depend upon both the magnitude and direction of the signal change. A conservative approach for inline systems is to sum all the time constants as equivalent dead time with the pure time delays and double the total to estimate the ultimate period.

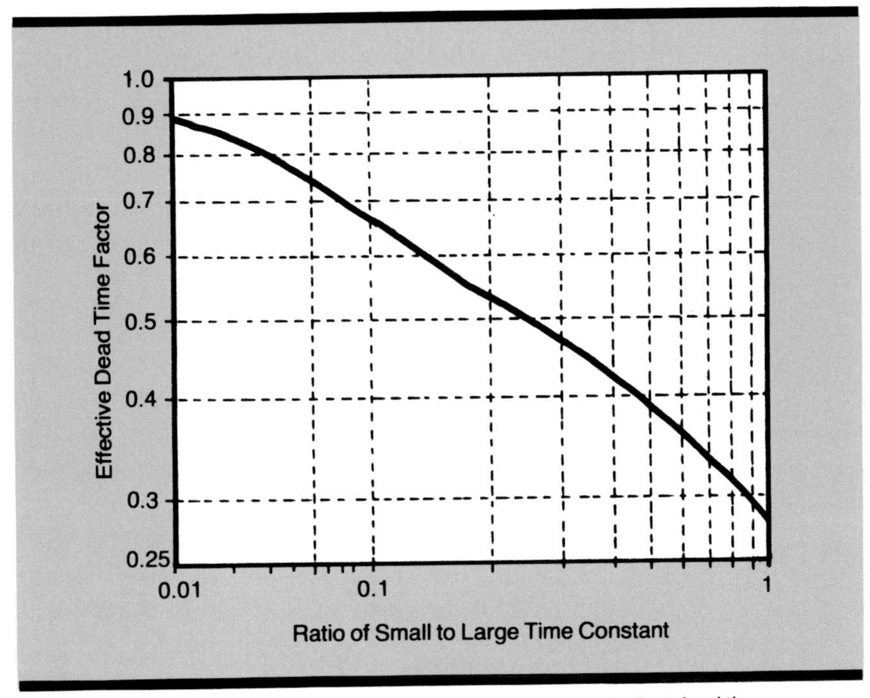

Fig. 9-1. The fraction of a small time constant converted to equivalent dead time approaches one as the ratio of the small to large time constant ratio approaches zero.

9-2. Controller Mode Settings

Ziegler and Nichols showed that the controller mode settings could be estimated from the ultimate proportional band and the ultimate period. The ultimate proportional band is the proportional band setting that caused the equal amplitude oscillations of the ultimate period. The ultimate proportional band can be estimated from the ultimate period, largest time constant, and steady-state gains for continuous pH control by Eqs. (9-3a) or (9-3b) and for batch pH control by Eqs. (9-4a) or (9-4b). The four steady-state open loop gains needed to calculate the ultimate proportional band are defined in Eqs. (9-5a) through (9-5e).

The valve gain is equal to the change in reagent flow divided by the percent change in controller output. For a linear valve, this gain is constant throughout the signal range. This gain is the same regardless of the type of control system.

The abscissa gain is equal to the change in abscissa divided by the change in reagent flow. F. G. Shinskey calls this gain the composition response gain (Ref. 4). For continuous pH control, this gain is equal to the inverse of the influent flow. For batch pH control, this gain is equal to the mass flow of pure reagent divided by total batch mass, which when integrated yields the weight fraction of reagent in the batch. This is the same integrator gain shown in Eq. (5-3) that was developed based on the assumptions that the reagent flow is constant, the batch is initially charged with the influent, and the reagent mass added is negligible compared to the batch mass.

The pH process gain is equal to the change in the titration curve ordinate divided by the change in the titration curve abscissa at the setpoint. F.G. Shinskey calls this gain the electrode gain in Ref. 5. This gain is equal to the slope of the line connecting the pH setpoint to the furthest excursion of pH. It is equal to the slope of the titration curve only for small changes in pH at the setpoint. Since the titration curve is nonlinear, this gain will depend upon the setpoint and the size of the disturbance. For a vertical well-mixed tank and low hysteresis control valve, the control band is tighter and therefore the change in gain is smaller. For a strong acid or strong base and a setpoint near the equivalence point, the gain is so high that the required proportional band exceeds the available proportional band (Ref. 6).

The measurement gain is equal to the change in percent controller input divided by the change in pH. This gain is constant throughout the signal range. The denominator of this gain is reagent flow instead of pH for a linear reagent demand controller.

The addition of signal characterization converts the nonlinear pH process gain to a unity gain. Of course, inaccuracies in the calculated reagent demand will cause deviations of this gain from unity. The multiplication of the output of the linear reagent demand controller by influent flow converts the abscissa gain to a unity gain for continuous pH control. The result is a fixed linear gain equal to the product of the valve gain and measurement gain. The proportional band no longer depends upon either the influent flow or the operating point. Consequently, controller tuning is simpler and doesn't require readjustment as long as the titration curve doesn't change shape.

For continuous pH control and $T_u < 4 * TC$ (vessels):

$$PB_u = \frac{100 * T_u * K_v * K_x * K_p * K_m}{2 \pi * TC} \quad (9\text{-}3a)$$

For continuous pH control and $T_u > 4 * TC$ (inline systems and ponds):

$$PB_u = 100 * K_v * K_x * K_p * K_m \quad (9\text{-}3b)$$

For batch pH control and $T_u < 4 * TC$ (vessels):

$$PB_u = \frac{100 * T_u^2 * K_v * K_x * K_p * K_m}{(2 \pi)^2 * TC} \quad (9\text{-}3c)$$

For batch pH control and $T > 4_u * TC$ (ponds):

$$PB_u = \frac{100 * T_u * K_v * K_x * K_p * K_m}{2 \pi} \quad (9\text{-}3d)$$

For continuous and batch pH and linear reagent demand control:

$$K_v = \frac{\Delta F_r}{\Delta O_c} \quad (9\text{-}4a)$$

For continuous pH control:

$$K_x = \frac{(\Delta F_r / F_i)}{\Delta F_r} = \frac{1}{F_i} \quad (9\text{-}4b)$$

For continuous linear reagent demand control (flow feedforward):

$$K_x = \frac{(\Delta F_r / F_i)}{\Delta F_r} * F_i = \frac{1}{F_i} * F_i = 1 \quad (9\text{-}4c)$$

For batch pH control:

$$K_x = \frac{\Delta X_b}{\Delta F_r} = K_i * \int dT = \frac{X_{rr} * F_r}{M_b} * \int dT \quad (9\text{-}4d)$$

For continuous pH control:

$$K_p = \frac{\Delta pH}{(\Delta F_r / F_i)} \tag{9-4e}$$

For batch pH control:

$$K_p = \frac{\Delta pH}{\Delta X_b} \tag{9-4f}$$

For continuous linear reagent demand control:

$$K_p = \frac{\Delta pH}{\Delta F_r} * \frac{\Delta F_r}{\Delta pH} = 1 \tag{9-4g}$$

For continuous or batch pH control:

$$K_m = \frac{\Delta I_c}{\Delta pH} \tag{9-4h}$$

For continuous linear reagent demand control:

$$K_m = \frac{\Delta I_c}{\Delta F_r} \tag{9-4i}$$

For continuous pH control:

$$K_v * K_x * K_p * K_m = \frac{\Delta F_r}{\Delta O_c} * \frac{(\Delta F_r / F_i)}{\Delta F_r} * \frac{\Delta pH}{(\Delta F_r / F_i)} * \frac{\Delta I_c}{\Delta pH} \tag{9-4j}$$

For continuous linear reagent demand control:

$$K_v * K_x * K_p * K_m = \frac{\Delta F_r}{\Delta O_c} * \frac{\Delta F_r}{\Delta F_r} * \frac{\Delta pH}{\Delta F_r} * \frac{\Delta F_r}{\Delta pH} * \frac{\Delta I_c}{\Delta F_r}$$

$$= \frac{\Delta F_r}{\Delta O_c} * \frac{\Delta I_c}{\Delta F_r} \tag{9-4k}$$

where:
ΔF_r = change in reagent flow (pph)

F_i = influent flow (pph)

I_c = controller input signal (%)

K_i = integrator gain (1/min)

K_m = measurement gain (%/pH) or (%/pph)

K_p = pH process (electrode) gain from titration curve (pH) or pH/pph)

K_v = valve gain (pph/%)

K_x = titration curve abscissa (composition) gain (1/pph)

M_b = total batch mass (lb)

O_c = controller output signal (%)

PB_u = ultimate proportional band (%)

ΔpH = change in pH

T_u = ultimate period (min)

TC = largest time constant in loop (min)

X_b = weight fraction of reagent in batch

X_{rr} = weight fraction of reagent in reagent stream

The product of the four gains is the open loop gain of the system which is the gain with the controller on manual. If all the gains are set up properly, the units in the numerators and denominators cancel out leaving a dimensionless open loop gain. Note that for continuous linear reagent control, the abscissa and pH process gains are unity so that just the valve and measurement gains are left. If the reagent demand range for the measurement signal is set equal to the control valve capacity, this gain product is also unity so that the open loop gain is one. For batch control, the gain product has the integrator gain units of inverse time that cancels out the additional time units in the equation for the ultimate proportional band. The open loop gain can be checked after startup by placing the controller in manual and making a 10% or more change in controller output and noting the final percent change in controller input for continuous control and the integration rate in percent per minute for batch control. The change in the controller input signal divided by the change in the output signal is the open loop gain. The open loop gain should be about one for a properly designed linear reagent demand controller.

The factors applied to the ultimate period and ultimate proportional band depend upon how many modes are used. The factors shown in Eqs. (9-5a) through (9-5f) were developed by Ziegler and Nichols (Ref. 7). This tuning method tends to provide a smaller peak error and smaller integrated error but a larger integrated absolute error than other methods. The actual value of the factors will change slightly with the type of controller (analog or digital) and the manufacturer and model number. Note that the proportional band for a three-mode controller reduces to simply the dead time to time constant ratio for a linear reagent demand controller on a well-mixed vessel since the open loop gain is one and the ultimate period is just slightly less than four times the dead time for a small dead time to time constant ratio. Equation (9-6), which shows this approximation, serves as a useful rule of thumb for tuning a well-designed control system. The dead time to time constant ratio is about equal to the turnover time divided by the residence time for such a system.

For a proportional-only controller:

$$PH = 1.8 * PB_u \tag{9-5a}$$

For a proportional plus integral (PI) controller:

$$PB = 2.2 * PB_u \tag{9-5b}$$

$$T_i = 0.8 * T_u \tag{9-5c}$$

For a proportional, integral, plus derivative (PID) controller:

$$PB = 1.7 * PB_u \tag{9-5d}$$

$$T_i = 0.5 * T_u \tag{9-5e}$$

$$T_d = 0.1 * T_u \tag{9-5f}$$

For a well-designed linear reagent controller on a well-mixed vessel:

$$PB = 1.7 * \frac{100 * 3.9 * TD}{2\pi * TC} = 100 * \frac{TD}{TC} \quad (9\text{-}6)$$

where:
PB = proportional band (%)

PB_u = ultimate proportional band (%)

T_d = derivative time (min)

T_i = integral time (inverse of reset setting) (min/repeat)

T_u = ultimate period (min)

TD = loop dead time (min)

TC = largest time constant in the loop (min)

Exercises

9-1. *What are the ultimate periods for continuous and batch pH control if the turnover time is 0.5 minute, the residence time is 10 minutes, the mixing time is 5 minutes, and the average electrode time constant is 0.1 minute?*

9-2. *What are the proportional, reset, and rate settings for continuous pH control for Exercice 9-1 data, the titration curve and control band in Fig. 6-1b, a valve capacity of 400 pph, an influent flow of 200 pph, and a pH transmitter range of 2 to 12 pH?*

References

[1]McMillan, G. K., *Tuning and Control Loop Performance*, ISA Monograph 4, 1983, pp. 42-43.
[2]*Ibid.*, pp. 48-49.
[3]*Ibid.*, pp. 34-49.
[4]Shinskey, F. G., *pH and pION Control in Process and Waste Streams*, John Wiley & Sons, 1973, pp. 133-134.
[5]*Ibid.*, pp. 140-141.
[6]McMillan, G. K., *Tuning and Control Loop Performance*, ISA Monograph 4, 1983, pp. 61-66.
[7]Ziegler, J. G., and Nichols, N. B., "Optimum Settings for Automatic Controllers," *ASME Transactions*, 1942, pp. 759-768.

Unit 10:
pH Control System Selection

UNIT 10

pH Control System Selection

This unit describes a logical method for selecting the electrode assemblies, control valves, vessels, and control system capable of being programmed for computer-aided design.

Learning Objectives—When you have completed this unit you should:

A. Recognize what data is needed for proper selection.

B. Know how to logically proceed through the selection decision process.

C. Appreciate what criteria is important for proper selection.

10-1. Instruments

The main performance criteria for proper selection of the electrode assemblies and control valves are reliability and accuracy for all pH loops and speed of response for inline pH loops. Reliability is achieved by identification of potential application problems and the minimization of them by instrument selection and the identification of possible operation interruption losses and the minimization of them by redundancy. Accuracy for the electrodes is achieved by the identification of the control band and by the assurance that the error is small in comparison by the minimization of the types of errors described in Unit 4-2. Accuracy for the control valves is achieved by the identification of the allowable reagent error from the control band and by the assurance that the hysteresis is less than this error by the use of low packing friction and high-performance positioners. In some applications, the variation in influent flow may dictate an even lower hysteresis to meet a rangeability requirement. Fast response is achieved for inline systems by the use of a high-velocity past the electrodes, by the elimination of piping between the control valve and the injection point, and by the use of a volume booster on the control valve.

Figure 10-1a shows the decision flowchart for electrode selection and Fig. 10-1b shows the decision flowchart for control valve selection. Each question requires a yes or no answer. The path to the next question depends upon answers to previous questions. The questions can be used to tabulate the data requirements. Note that the flowcharts provide the general types of

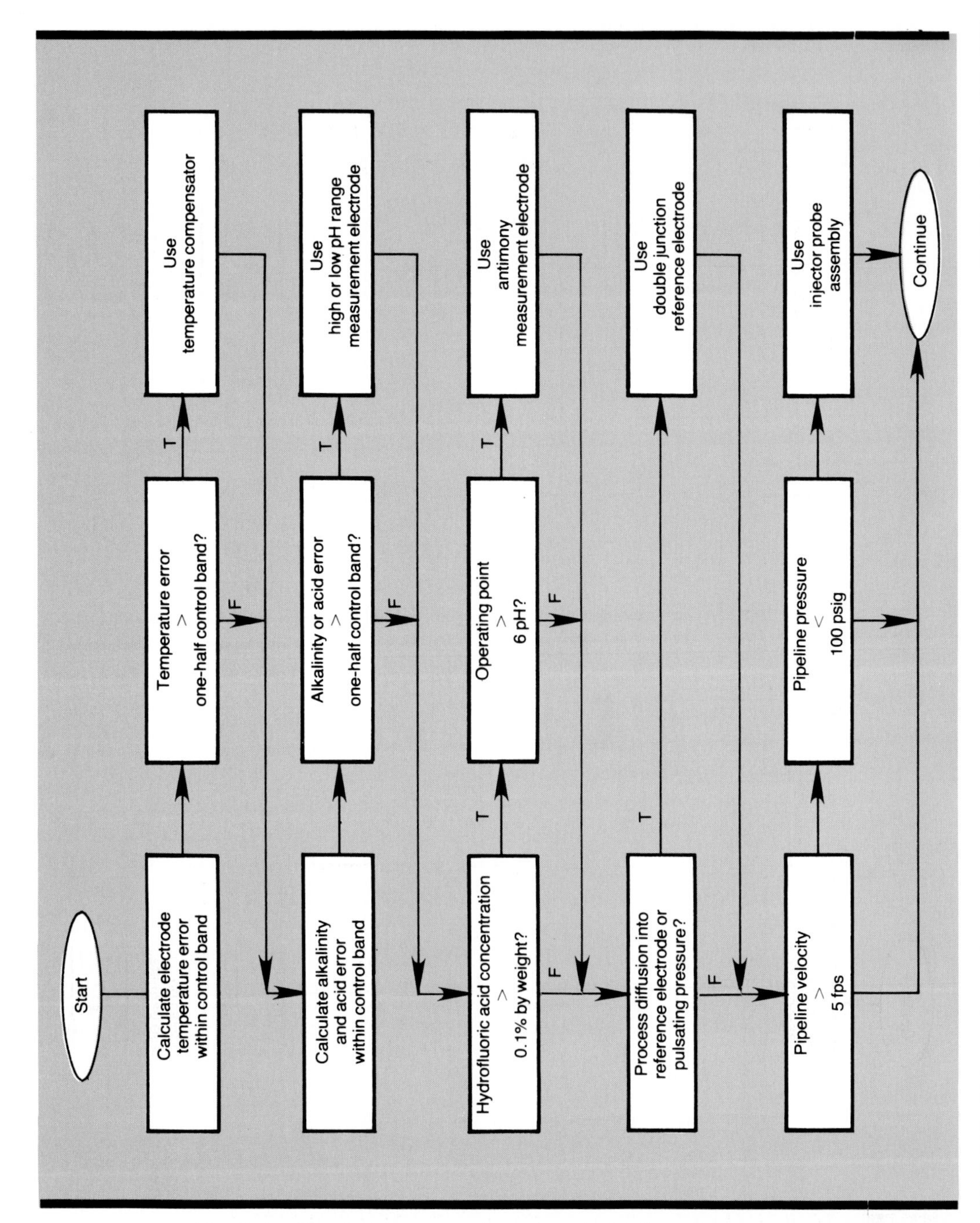

Fig. 10-1a. The electrode assembly must meet reliability and accuracy requirements for all pH loops and the speed of response requirement for inline pH loops.

instruments required. The instrument engineer must still specify the details such as the materials of construction, connections, and range or capacity based on specific application data.

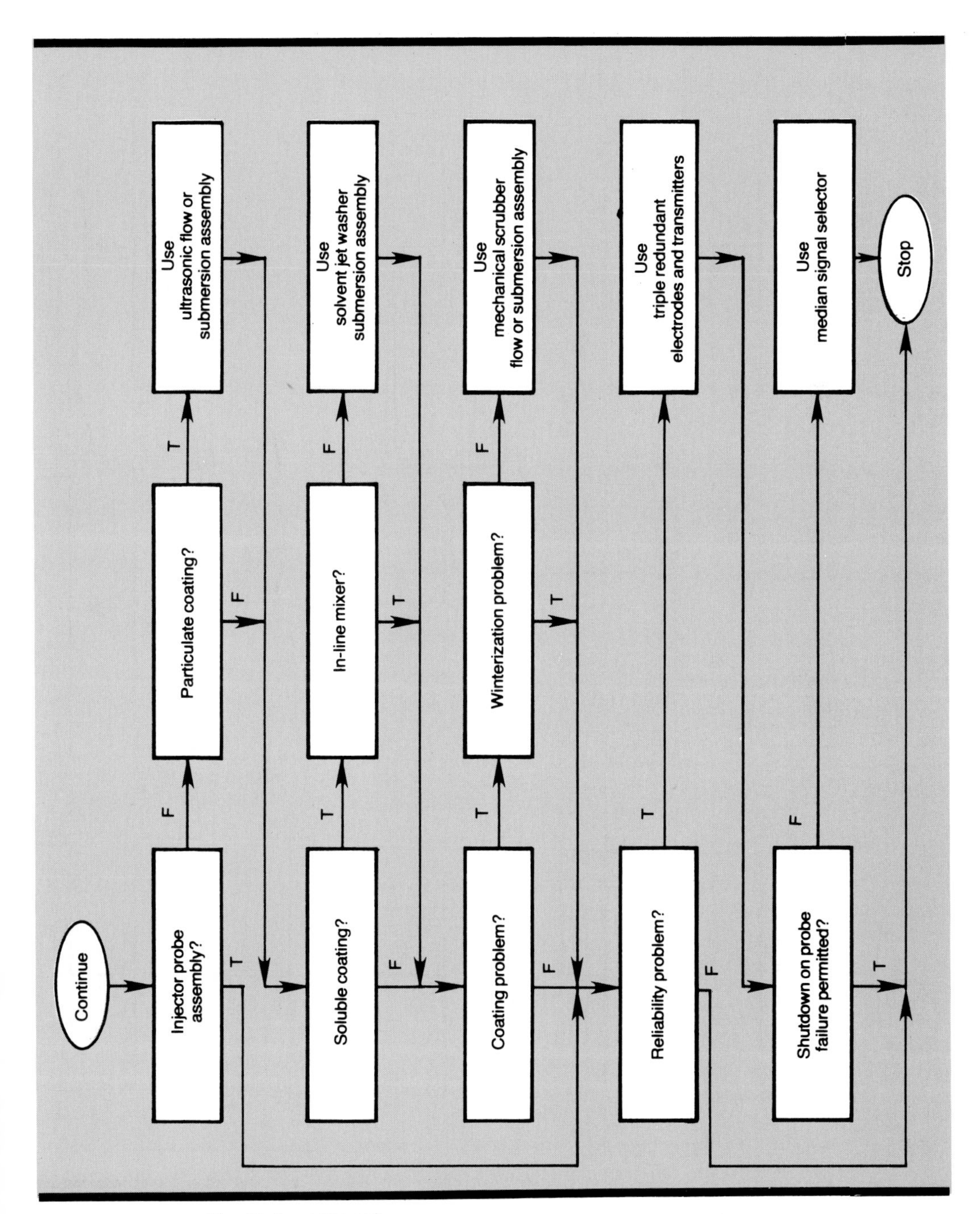

Fig. 10-1a continued.

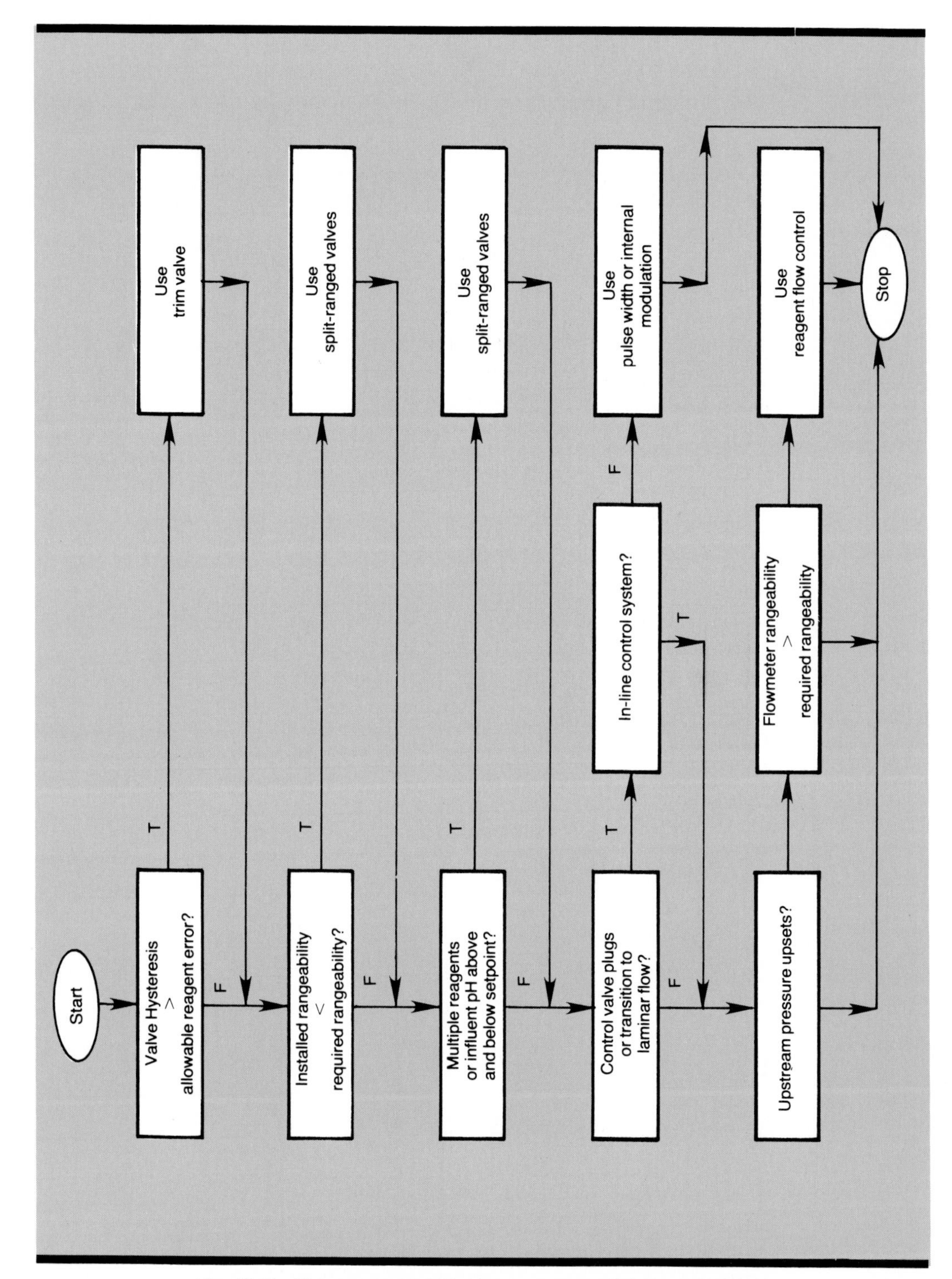

Fig. 10-1b. The reagent control valve must meet reliability and accuracy requirements for all pH loops and the speed of response requirement for inline pH loops.

10-2. Vessels and Control Systems

The main performance criteria for the selection of vessels and control systems are the peak and integrated errors. The peak error is the maximum excursion of the pH from setpoint after a disturbance. Ideally, the peak error should be within the control band for all conceivable disturbances. However, it is inappropriate to spend extra capital to protect against a peak error that lasts only a few seconds for a disturbance that seldom occurs. The extreme operating conditions associated with such unusual disturbances may require an extra vessel or a control valve rangeability that necessitates multiple reagent valves and the associated controls that increase the complexity and difficulty of day-to-day operation. The filtering effect of downstream volumes on the peak error is often ignored despite significant potential attenuation. The attenuation of the peak error can be estimated by translating the peak pH error to a peak reagent error via the titration curve, using Eq. (10-1a) to find the filtered reagent error, and translating the filtered reagent error back to a filtered pH error via the titration curve. The loop period in Eq. (10-1a) can be approximated as the ultimate period for PID control and about 1.5 times the ultimate period for PI control for the tuning method described in Sec. 9-2. If the proportional band is larger than required, the peak error and the control loop period will both be increased. The double effect on the filtered error can cause the filtered error to exceed the control band.

$$E_f = \frac{E_p * T_o}{TC_p * 2\pi} \tag{10-1a}$$

where:

E_f = filtered reagent error (pph)

E_p = peak reagent error (pph)

T_o = control loop period (minutes)

TC_p = equipment time constant (minutes)

The peak reagent error can be estimated by Eq. (10-1b) for continuous pH control. If the proportional band is small compared to 100% times the open loop gain, the proportional band term can be dropped from the denominator to yield Eq. (10-1c). This equation can also be used for batch pH control. For small dead time to time constant ratios and continuous pH control, Eq. (10-1c) reduces to Eq. (10-1d) that serves as a useful rule of thumb. The substitution of Eqs. (5-1a) and (5-2a) for a well-mixed vessel yields Eq. (10-1e) that shows the peak error is independent of vessel size if the input flow is small compared to the agitator pumping rate and the entire volume is backmixed. The horsepower increases with the fourth power of the diameter of the impeller so that it is extremely expensive to provide complete backmixing of large vessels.

$$E_p = \frac{K * PB}{(100 * K_v * K_x * K_p * K_m + PB)} * E_o \qquad (10\text{-}1b)$$

$$E_p = \frac{K * PB}{100 * K_v * K_x * K_p * K_m} * E_o \qquad (10\text{-}1c)$$

$$E_p = \frac{TD}{TC} * E_o \qquad (10\text{-}1d)$$

$$E_p = \frac{Q_t}{Q_a} * E_o \qquad (10\text{-}1e)$$

$$C_d = e^{\frac{(-4 * TC_c)}{T_u}} \qquad (10\text{-}2)$$

where:

C_d = correction factor for slowness of disturbance

E_p = peak reagent error (pph)

E_o = open loop reagent error (pph)

K = factor that depends on controller type and number of modes (typically K = 1.1)

K_m = measurement gain (%/pH) or (%/pph)

K_p = pH process (electrode) gain from titration curve (pH) or (pH/pph)

K_v = valve gain (pph/%)

K_x = titration curve abscissa (composition) gain (1/pph)

PB = controller proportional band (%)

TC = largest time constant in the loop (minutes)

TD = total loop dead time (minutes)

TC_d = disturbance time constant

T_u = ultimate period of loop (minutes)

Q_a = agitator pumping rate (gpm)

Q_t = total input flow (gpm)

An increase in the open loop gain product does not change the fraction of the open loop reagent error seen as the peak reagent error because the proportional band is proportional to this gain product so that the gain product in the denominator is cancelled out by gain product in the numerator in Eq. (10-1c) (Ref. 1). However, an increase in the abscissa gain increases the open loop reagent error for a given reagent or influent disturbance and an increase in the pH process gain increases the peak pH error for a given peak reagent error. The errors predicted by these equations are for step disturbances. If the disturbance time constant is small compared to the loop dead time, the disturbance can be considered as equivalent to a step disturbance. The effect of the disturbance time constant can be roughly estimated by multiplying the results of Eqs. (10-1b) through (10-1d) by the correction factor from Eq. (10-2). As the disturbance time constant approaches zero, the correction factor approaches unity. As the time constant approaches infinity, the correction factor approaches zero (Ref. 2). The control band for such extremely large disturbance time constants is determined by the pH electrode error and the control valve hysteresis. It is important to remember that these two errors cannot be used as open loop errors in Eqs. (10-1b) through (10-1d) because control action cannot reduce them.

The integrated error is the total positive area minus the total negative area between the setpoint and measurement. Equation (10-3a) can be used to estimate this error in reagent units as long as the loop oscillations are decaying (Ref. 3). The integrated error provides a measure of the total amount of the excess reagent in the effluent. The time units for the mass flow for the open loop reagent error is cancelled by the time period for integration so that the result is total pounds of excess reagent. The effect of a disturbance on the pH of a downstream holding tank can be estimated by dividing this integrated error by the total mass of the effluent and translating this weight fraction to a pH via the titration curve. For small dead time to time constant ratios and pH control, Eq. (10-3a) reduces to Eq. (10-3b), which shows that the integrated error is proportional to the dead time squared (Ref. 4). The substitution into Eq. (10-3b) of Eq. (10-1e) yields Eq. (10-3c) that shows the integrated error is proportional to the vessel volume for well-mixed vessels. The integral time is proportional to the ultimate period that is proportional to the vessel size for a fixed agitator pumping rate. An extremely large vessel will require an integral time setting larger (reset setting smaller) than that available on industrial controllers. The result is a continuous reset cycle that can be dissipated by increasing the proportional band for continuous pH control. It cannot be dissipated for batch pH control due to process non-self-regulation. If the agitator pumping rate was scaled up with the vessel size, the integrated error would stay the same and the peak error would be less. However, the horsepower requirement quickly gets out of hand. Unit 4 showed that unagitated vertical vessels still have a major portion of their residence size available as an equipment time constant for the attenuation of pH oscillations. Thus large vessels improve the system performance if the control loops are upstream or downstream but not on the vessel, unless the agitation flow is scaled up with the vessel volume. If the large vessel is upstream of the loop, the reagent usage is decreased because the pH control loop does not have to react to short-term pH excursions. The greatest savings in reagent occurs when the upstream volume averages out alternate swings of the influent from an acidic to basic condition.

$$E_i = \frac{PB * T_i}{100 * K_v * K_x * K_p * K_m} * E_o \tag{10-3a}$$

$$E_i = \frac{TD^2}{TC} * E_o \tag{10-3b}$$

$$E_i = \frac{Q_t * V}{Q_a^2} * E_o \tag{10-3c}$$

where:

E_i = integrated reagent error (lbs)

E_o = open loop reagent error (lbs/min)

PB = proportional band

TC = largest time constant in the loop (minutes)

TD = total loop dead time (minutes)

Q_a = agitator pumping rate (gpm)

Q_t = total input flow (gpm)

V = vessel working volume (gallons)

Figure 10-2 is a plot of the response to a step influent disturbance of a composition control loop on a static mixer, well-mixed vessel, and poorly mixed vessel. The plot shows the relative performance in terms of peak and integrated errors of a control loop on the three types of equipment. The static mixer loop has the greatest peak error but the smallest integrated error. The well-mixed tank loop has the smallest peak error but an intermediate integrated error. The poorly mixed tank has an intermediate peak error but the largest integrated error. Thus, the static mixer is a viable alternative to a well-mixed tank if the following conditions described in other units are satisfied:

Conditions for Successful Use of a Static Mixer

(1) A downstream volume exists to attenuate the peak error.

(2) The setpoint is on a relatively flat portion of the titration curve.

(3) The offset between filtered pH and setpoint is acceptable.

(4) The control valve is close-nippled into the injection point.

(5) The number of injections or velocity is enough to insure premixing.

(6) An injector type of electrode assembly is used.

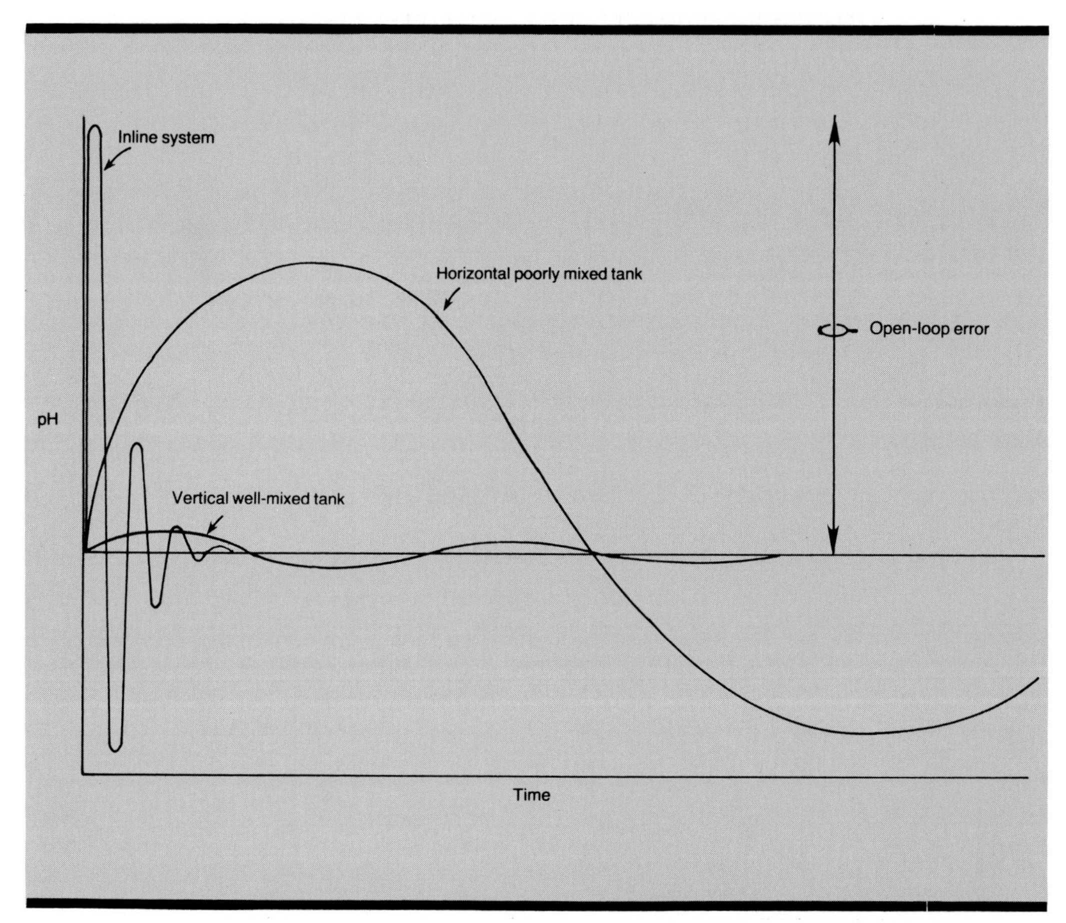

Fig. 10-2. The static mixer loop has the smallest integrated error, the well-mixed tank has the smallest peak error, and the poorly mixed tank has the largest integrated error.

An existing unagitated storage tank can be used in conjunction with an inline pH control loop to provide tight pH control if the conditions for a static mixer are satisfied. The reagent and influent streams are injected into the suction of the circulation pump. Flow feedforward control should be used if the influent flow is not flow controlled. An injector type of pH probe with a double junction reference electrode is installed at the discharge of the pump as shown in Fig. 10-3. The centrifugal pump acts as an efficient low-dead time dynamic mixer. The effluent stream is discharged by level control from the tank. The tank acts as an excellent filter of loop oscillations and acts as a moderator of influent disturbances since the circulation flow is large compared to the influent flow. The mass of fluid in the tank is extremely large compared to the integrated reagent error from inline pH control so that the weight fraction of reagent error in the tank is very small. Field startup results show a virtual straight line for the effluent pH during upsets, which is previously unheard of in pH control. This system responds quickly during startup because the influent is immediately sensed. If the influent was added to the tank, influent disturbances would be filtered, but a large overshoot might occur on startup because a large amount of influent would have been added to the tank before it was detected by the pH electrode.

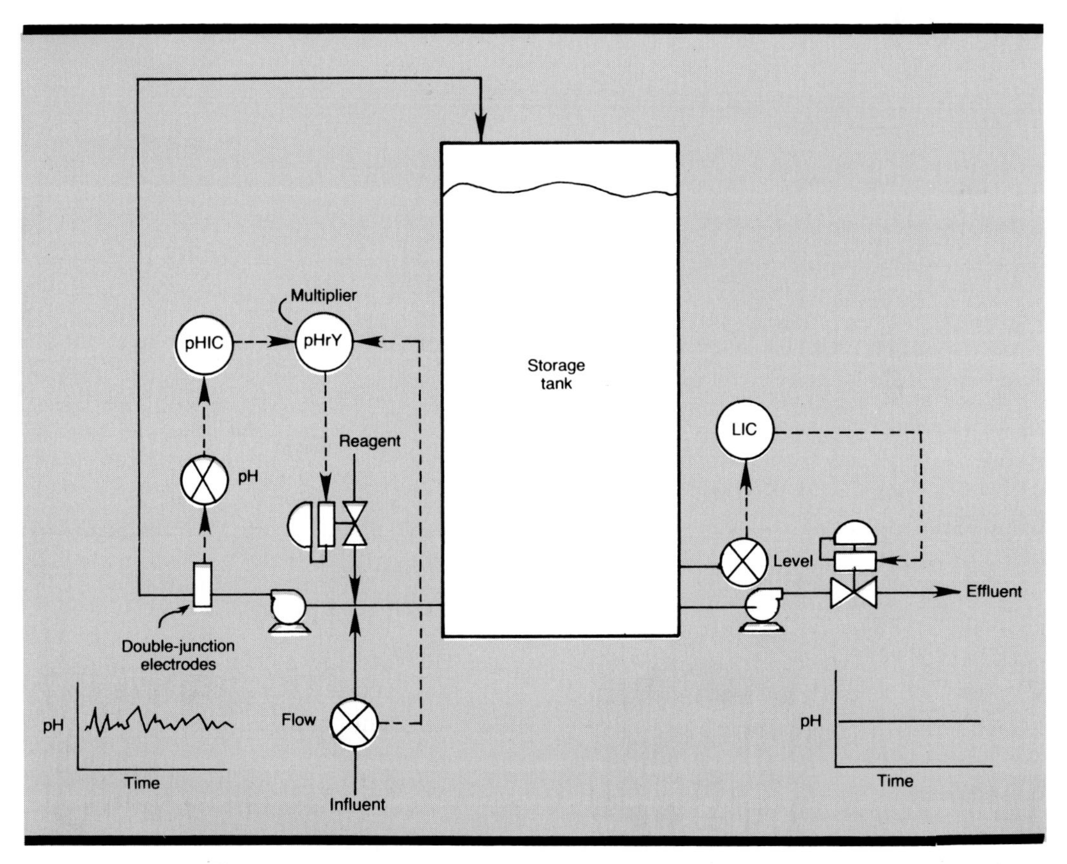

Fig. 10-3. A storage tank, circulation pump, and in-line pH system can be used to provide tight pH control.

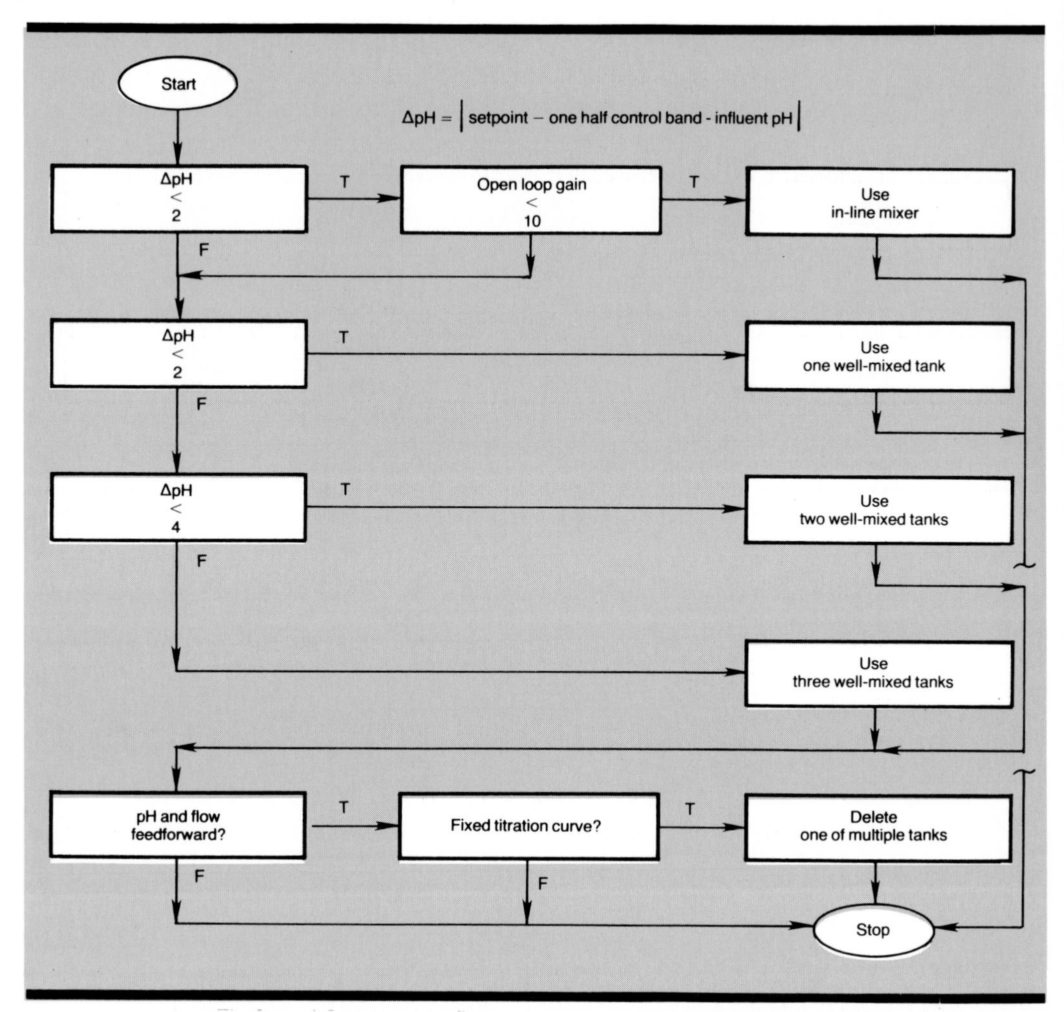

Fig. 10-4a. A conservative short cut method can be used to estimate the number of vessels and loops in series without any disturbance or loop dynamic information.

Figure 10-4a shows a conservative shortcut method of estimating the number and type of equipment and control loops. It does not require information on the dynamics and magnitude of the disturbances, which is typically difficult to get. While it does not use a titration curve, a curve is still needed for control valve selection and controller tuning. The method is based on the deviation of the influent pH from the setpoint. A large deviation is characteristic of a titration curve with a steep slope. To this extent, the method does infer the shape of the curve.

The use of feedforward pH and flow control can eliminate the need for an additional vessel and feedback loop if the influent pH measurement is accurate enough and the titration curve shape is constant enough. The use of linear reagent demand feedback

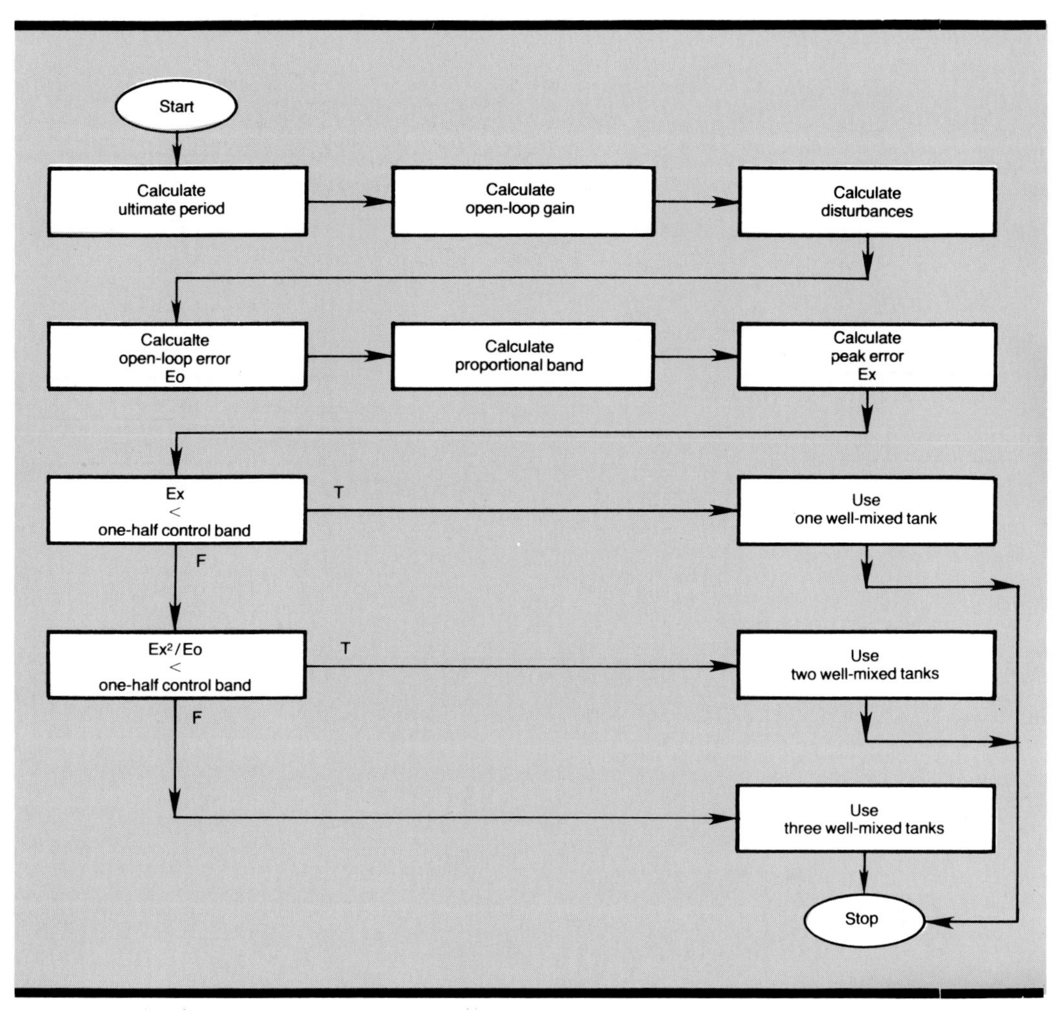

Fig. 10-4b. A more rigorous method can be used to estimate the number of vessels and loops in series if the disturbance and loop dynamics are known.

control will improve the performance of the loop and can eliminate the need of an additional vessel for applications where the additional vessel tank requirement was borderline.

Figure 10-4b shows a more rigorous method of estimating the number and type of equipment and control loops. It requires information on the dynamics and magnitude of the disturbances. The basic strategy is to calculate the peak reagent error for a given open loop reagent error for each vessel. The open loop reagent error of the next vessel is the peak reagent error of the previous vessel. The peak reagent error is translated to a peak pH error at the effluent of each tank to check if it is reduced enough to be within a specified control band. The iteration continues until the peak error is small enough. It is important to realize that

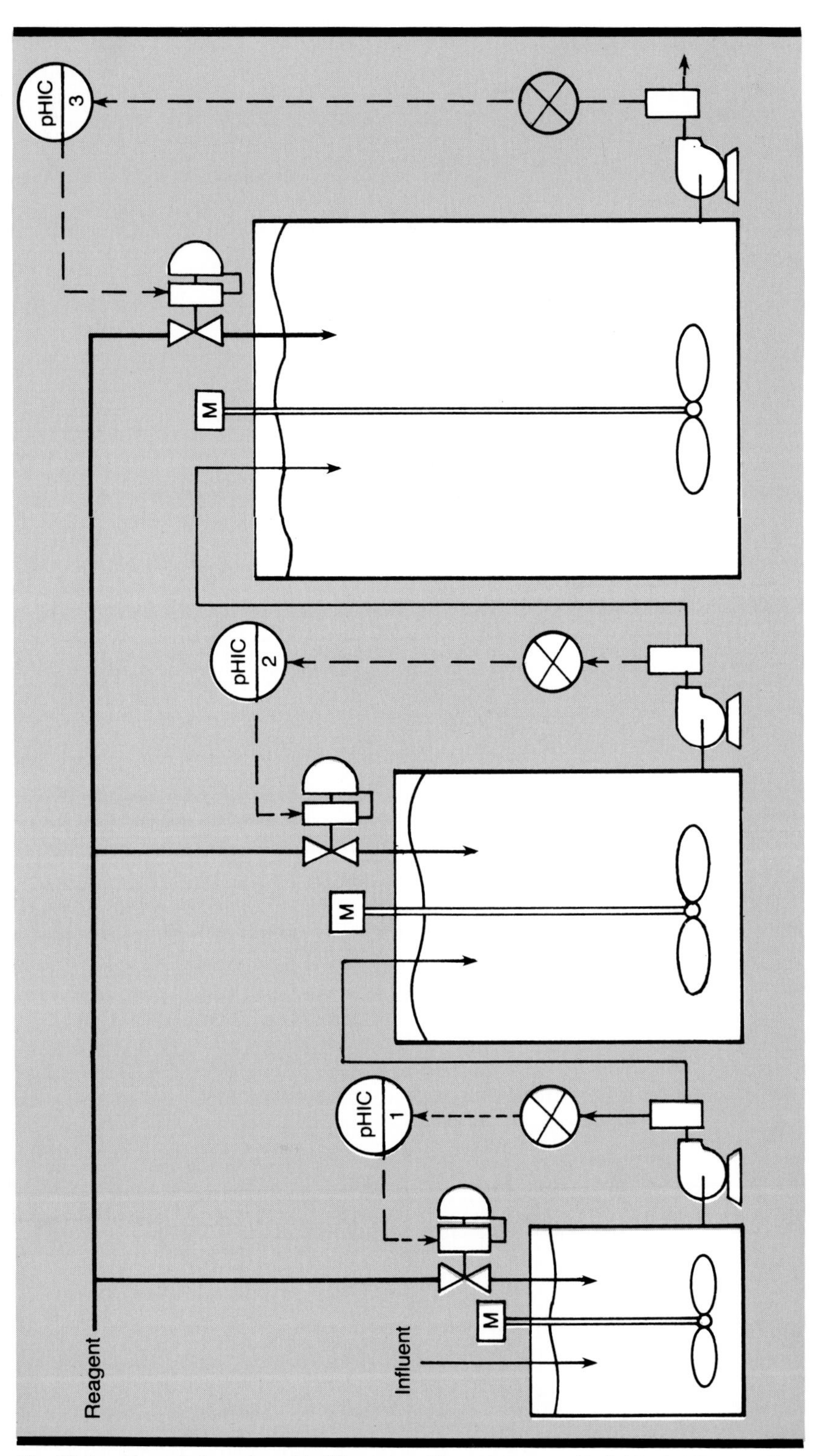

Fig. 10-5a. Vessels in series with different working volumes and individual pH loops will greatly decrease the final pH error.

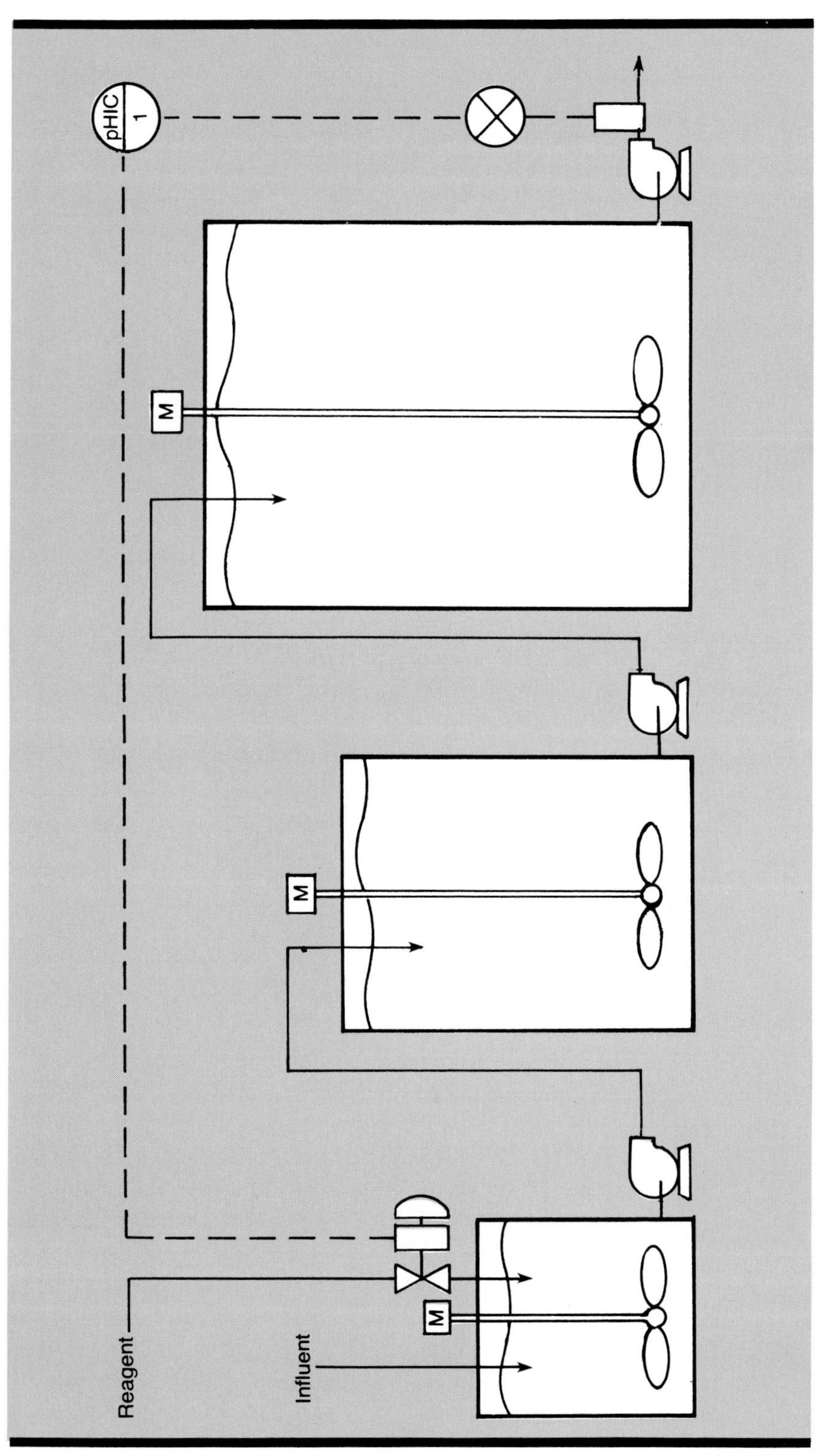

Fig. 10-5b. A single overall pH loop around vessels in series will greatly increase the final pH error.

individual pH control loops must be installed on each vessel as shown in Fig. 10-5a. If one overall pH control loop is installed for the vessels in series, as shown in Fig. 10-5b, the performance will be much worse than that for a single vessel (Ref. 5). The equipment time constants of the additional vessels between the pH controller and control valve are converted into an equivalent dead time per Fig. 9-1. Each vessel's working volume should be different so that the equipment dead time, and hence loop ultimate period, is different. Equal ultimate periods may cause disturbances to resonate if the disturbance becomes in phase with loop oscillations in successive tanks.

The use of multiple vessels in series facilitates the use of multiple reagent valves for a given reagent without split ranging or valve position control. The first vessel should have the largest valve and the last vessel should have the smallest or trim control valve. The actual capacity of each valve depends upon the distance in reagent units along the abscissa of the titration curve from the influent pH to the setpoint pH. The influent pH of each successive vessel is the setpoint pH of the previous vessel. The reagent peak error should be added to the distance in reagent units to account for influent pH variability. The control valve rangeability dictates how far apart the setpoints can be. A rangeability of 100:1, which corresponds to one percent hysteresis, is a reasonable guess for a linear globe valve with teflon packing and a positioner.

The method shown in Fig. 10-4b is for a linear control loop. A linear reagent demand controller for a fixed titration curve can achieve the estimated performance. The performance of a pH loop can be considerably worse due to the nonlinearity of the titration curve. The problem is self-aggravating in that a larger control band causes a larger nonlinearity that causes a larger control band and so on. There is no clear-cut method of estimating the deterioration in loop performance due to the nonlinearity. If the controller proportional band was increased to stabilize the pH loop for a setpoint on the steep portion of the titration curve, this setting should be used to estimate the error for excursions outside of the band. Another method used by the author estimates the change in dead time or time constant due to the nonlinearity of the titration curve. Whereas the dead time and time constant of the composition response may be fixed by the equipment turnover time and residence time, the pH response seen by the pH controller has a different effective dead time or time constant. If the setpoint is on the steep portion of the titration curve and a disturbance causes an excursion to the flat portion, the time it takes to reach 63% of its final pH (one time constant) is much shorter. The dead time to time constant ratio seen by the controller is much bigger. For example, a 19 minute composition time constant is reduced to a 0.04 minute pH time constant for a strong acid and strong base system (Ref. 6). The problem this poses for the control loop can be visualized by remembering how quickly the pH passes through the equivalence point during a titration or during pH control. In fact, some engineers try to install an electronic filter to slow down the passage through the equivalence point. However, the actual pH peak error is increased even though the attenuated controller input peak error may be decreased. The loop period is also increased and hence the integrated error is increased. An electronic filter should only be used to attenuate measurement noise. The time constant seen by the pH controller can be estimated by marking off the open loop reagent error along the abscissa and translating this to an open loop pH error on the ordinate of the titration curve. The 63% point of the open loop pH response is translated to a smaller percent of the open loop response and the effective time constant is solved for by Eq. (10-4). Figure 10-6 illustrates the graphical application of the method.

$$TC_e = -\ln\left[1 - \frac{E_e}{E_o}\right] * TC \tag{10-4}$$

where:

E_o = the open loop reagent error (pph)

E_e = the effective distance to 63% of the open loop pH error (pph)

TC = largest time constant in the loop (minutes)

TC_e = effective time constant seen by the pH controller (minutes)

The linear reagent demand controller slows down the controller input signal as it passes through the equivalence point. The linear reagent demand controller restores the effective time constant seen by the pH controller back to the composition time constant so that the full potential benefit from the capital and energy cost of a well-mixed vessel can be achieved.

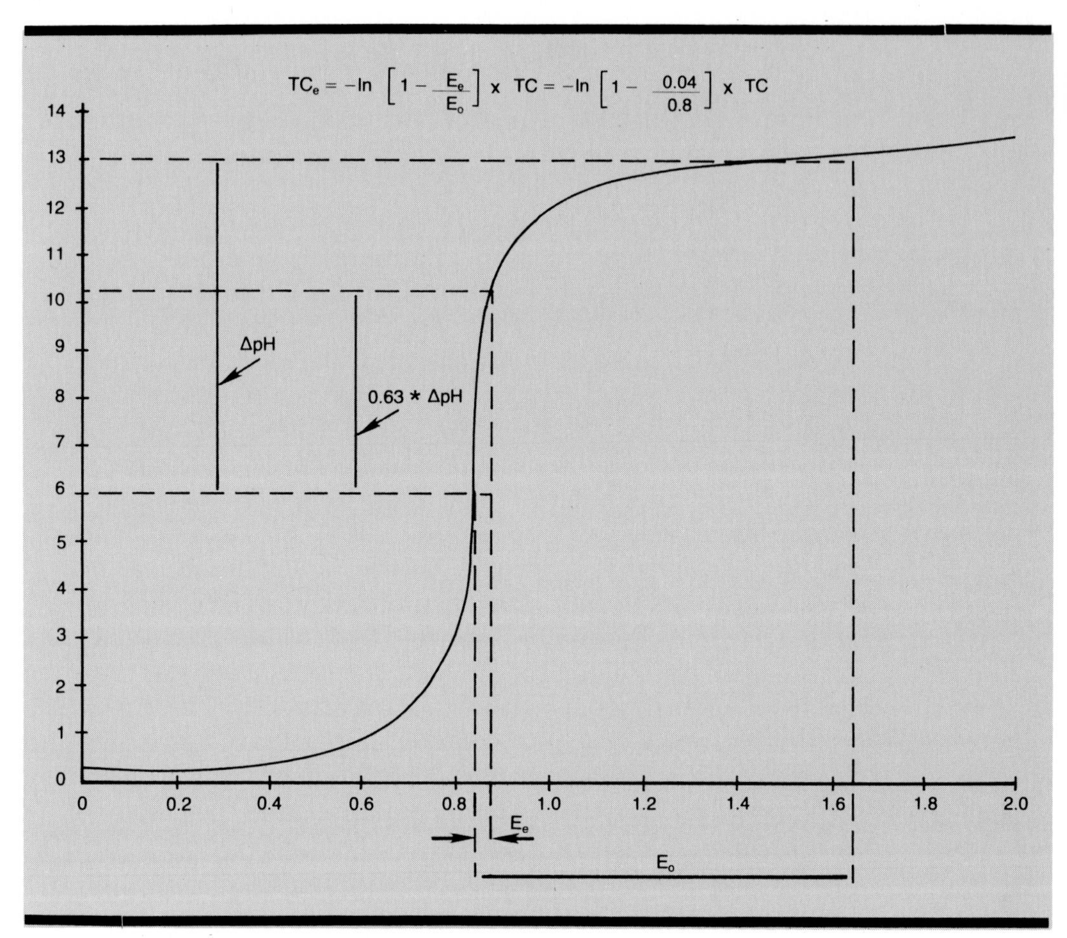

Fig. 10-6. The nonlinearity of the titration curve decreases the effective time constant seen by the pH controller for a setpoint on the steep portion of the titration curve.

Exercises

10-1. *If the setpoint is 5 pH, the effluent temperature is 40°C, and the control band is 2 pH, should a temperature compensator be used?*

10-2. *If the electrode coating is soluble and gravity overflow is used for level control, what type of electrode assembly should be used?*

10-3. *What is the least expensive method of reducing the plugging of a reagent control valve for concentrated sulfuric acid for vessel pH control?*

10-4. *What are the vessel and control system options for a control band of 1 pH, a setpoint of 7 pH, an influent at 4 pH, and a fixed titration curve?*

10-5. *What are the vessel and control system options for the conditions of Exercise 10-4, a turnover time of 0.5 minutes, a residence time of 10 minutes, a titration curve shown in Fig. 6-1b, an influent flow of 200 pH, and a step disturbance that causes an open loop reagent error of 50 pph?*

References

[1]McMillan, G. K., *Tuning and Control Loop Performance*, ISA Monograph 4, 1983, pp. 28-29.
[2]*Ibid*., pp. 157-158.
[3]*Ibid*., pp. 24-28.
[4]*Ibid*., pp. 45, 46.
[5]*Ibid*., pp. 63-66.
[6]*Ibid*., pp. 175-177.

Unit 11: pH Control System Checkout

John C. Pfeiffer, P.E.
Pfeiffer Engineering Co.
560 Garden Drive
Louisville, Kentucky 40206

UNIT 11

pH Control System Checkout

This unit describes a general calibration procedure and some of the problems to watch for during installation, water batching, and commissioning.

Learning Objectives—When you have completed this unit you should:

A. Understand the use of the different calibration adjustments.

B. Recognize common installation pitfalls.

C. Recognize potential problems during water batching and commissioning.

11-1. Calibration Check

The pH meter and transmitter calibration should be checked before installation to verify integrity as received and again before startup to verify integrity as installed. The pH meter and transmitter may not have all the calibration adjustments shown in Eqs. (4-5a) and (4-5b) and may use different names for the ones it does have. The meter zero adjustment may be omitted or called a bias or offset adjustment. The standardization adjustment is seldom omitted but may be called the input zero adjustment. The meter span adjustment may be omitted or called a slope adjustment. The transmitter output span and zero adjustments are seldom omitted, but the zero adjustment may be called an elevation adjustment.

The standardization and meter span adjustments interact for any pH range that does not start at the isopotential point. Thus, these adjustments must be repeated until desired accuracy is achieved. This interaction can be eliminated by the use of the meter zero to shift the isopotential point to the start of the pH range. This method is also effective in compensating for the nonlinearity in a pH range above or below 7 pH (Ref. 1). Figure 11-1 shows how the vertical downward shift of the isopotential point by the meter zero adjustment improves the match of the meter line to the electrode curve. The following general calibration procedure illustrates the use of these and the other calibration adjustments.

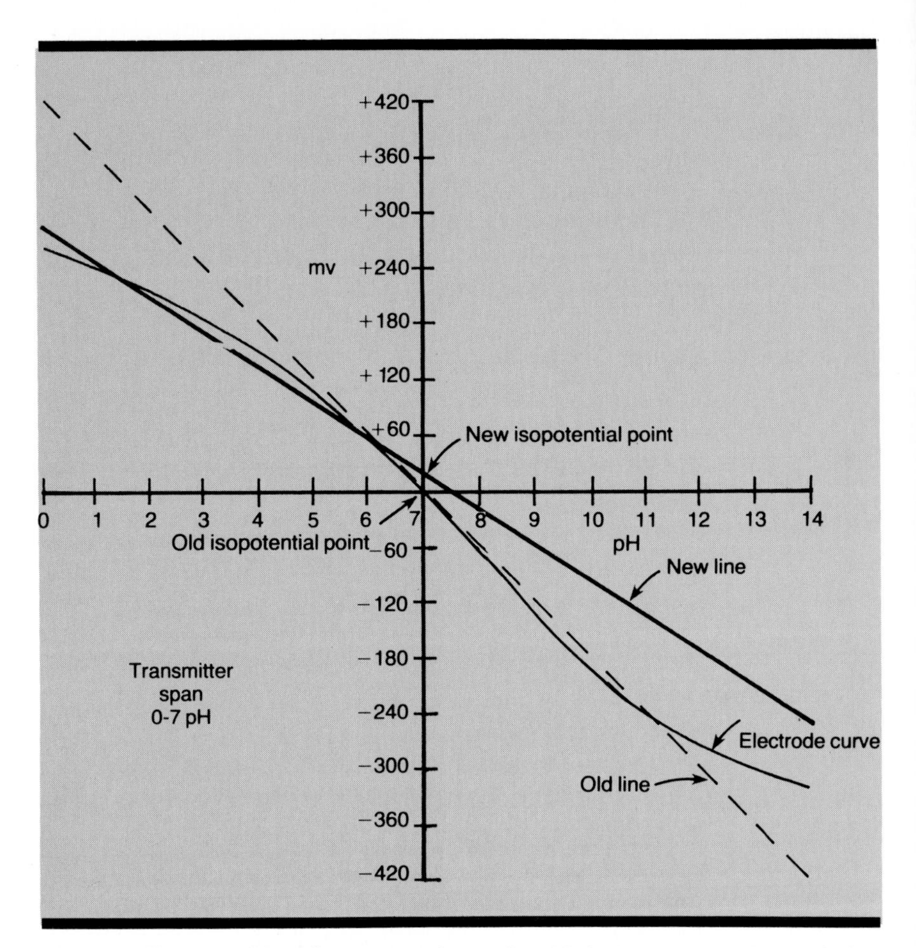

Fig. 11-1. The meter zero adjustment can be used to shift the isopotential point to help the meter calibration line match the electrode curve.

General Calibration Procedure

(1) If there is no automatic temperature compensation, set the manual temperature compensator dial to the buffer temperature.

(2) Switch the meter to "standby" (shorting strap is connected from the measurement to reference electrode input terminals).

(3) Adjust the meter zero to give a 7 pH reading or the pH reading at the low end of a nonlinear pH range.

(4) Switch the meter to "operate" or "run" (shorting strap is removed from the input terminals).

(5) Immerse the electrodes in a buffer with a pH that matches the low end of the pH range.

(6) Adjust the standardization to make the pH meter read the buffer pH.

(7) Remove, rinse electrodes with distilled water and pat dry, or rinse electrodes with the next buffer solution.

(8) Immerse the electrodes in a buffer with a pH that matches the high end of the pH range.

(9) Adjust the meter span to make the pH meter read the buffer pH.

(10) Repeat steps 5 through 9 with a rinse before each immersion until the desired accuracy is obtained.

(11) Immerse the electrodes in a buffer with a pH that matches the low end of the pH range or use an electrode simulator to generate the same pH.

(12) Adjust the transmitter zero to make the current output equal 4 madc or the voltage output equal 1 vdc.

(13) Remove, rinse electrodes with distilled water and pat dry, or rinse electrodes with the next buffer solution if an electrode simulator is not used.

(14) Immerse the electrodes in a buffer with a pH that matches the high end of the pH range or use an electrode simulator to generate the same pH.

(15) Adjust the transmitter span to make the current output equal 20 madc or the voltage output equal 5 vdc.

The electrodes should not be wiped dry because the static charge created could cause an error. The buffer value stated on the bottles is for room temperature. The buffer pH should be corrected for the ambient temperature during field calibration. For example, a phosphate buffer solution of 6.865 pH at 25°C will actually be 6.923 pH at 10°C.

11-2. Installation Check

The electrodes should not be installed until startup to prevent dehydration or damage during construction and flushing of lines. The electrode connections, conduit, preamp, and transmitter are installed soon after the equipment and piping is installed. The injector pH probe can use installation details similar to that used for filled thermowells to obtain the required insertion length. The installation should be reviewed for the following pitfalls:

Measurement Installation Pitfalls

(1) Transmitter housing is not purged. (Moisture accumulated on terminals.)

(2) The cable to the transmitter and preamp is not shielded. (Signal picks up electrical noise.)

(3) The cable's shield to the transmitter and preamp is grounded. (Ground loop is created.)

(4) The preamp or transmitter is removed from an ungrounded fiberglass enclosure and put in a grounded metal enclosure. (Ground loop is created.)

(5) The sample line is not winterized. (Sample line freezes during water batching or when sample flow is shut off.)

(6) The electrode tips are not pointing down. (Air bubble forms inside the electrode tips.)

(7) The electrode tips are in the top of a line or near the surface of a vessel. (Air bubbles form in the effluent near the electrode tips.)

(8) The slot in the sheath of an injector probe is pointed into the flow for a slurry stream. (Electrode abrasion develops.)

(9) The slot in the sheath of an injector probe is not pointed into the flow for a fouling stream. (Electrode coating develops.)

(10) The electrodes are not kept wetted for batch or intermittent operations. (Dehydration of electrode occurs between batches.)

(11) The injector or submersion electrode assembly has an overhead obstruction. (Insufficient clearance is available for removal of the electrode assembly.)

(12) Submersion electrode assembly is near bottom of vessel and is not supported. (Assembly breaks due to whipping action from agitation.)

(13) Submersion electrode assembly is installed in a stilling well. (Response is sluggish and coating develops due to insufficient velocity past the electrodes.

11-3. Water Batching

The electrodes should, in general, not be installed during water batching because the lines may not be clean and the measurement is meaningless in most cases. Sometimes the control system is tested by attempting pH control of the water by reagent addition. The results are usually poor. The water stream has little buffering. A small amount of reagent addition causes a large pH change and the reagent control valve capacity is too large to prevent large oscillations. Tuning the controller for these conditions is a waste of time. Operations may commit their time and maintenance to solving a performance problem that only exists during water batching. At the very least, confidence in the pH control system is lost. The key point is that the control system was designed for a titration curve and set of operating conditions that exists during production, not during water batching.

11-4. Commissioning

It is important that the controller modes be adjusted based on intelligent guesses and the controller be on automatic before the influent flow starts. Otherwise, a large amount of untreated influent can accumulate and cause equipment corrosion and environmental violations due to an excessively long loop recovery time. The logarithmic relationship of pH means that a 3 pH deviation of pH corresponds to a weight fraction of excess influent that is 1000 times too large. A pH feedforward or linear reagent demand feedback control system can significantly reduce the loop recovery time for startup. A flow feedforward control system may not reduce the recovery time because the flow feedforward signal may be multiplied by a zero from the feedback controller output.

If the open loop or reaction curve method of controller tuning is used, the time constant that is graphically measured is not the composition time constant but the effective pH time constant seen by pH controller due to the nonlinearity of the titration curve. This time constant will change with the size of the change in the manual output. The open loop gain can be checked by dividing the final change of the controller input in percent by the change of the controller output in percent. If the ultimate oscillation method of controller tuning is used, the amplitude of the first set of steady oscillations is a good indicator of where the breakpoints should be set for a nonlinear or breakpoint pH controller. The proportional band setting divided by the notch gain setting should equal the proportional band estimated from the ultimate proportional band. Regardless of the type of method used for on-line tuning, the starting point should be the pH setpoint due to operating point nonlinearity.

Exercises

11-1. *Why would the meter zero adjustment be helpful for a range of 8 to 14 pH?*

11-2. *Why is a buffer solution instead of a process sample used to calibrate the pH meter?*

11-3. *What might happen if the steam was left on to a winterized sample line during warm weather?*

11-4. *What can be done to install an injector probe that requires an insertion depth of 7 inches to a 2 inch pipeline?*

11-5. *Why is electrode breakage a problem during water batching?*

11-6. *Why does pH feedforward reduce the recovery time for startup and flow feedforward does not?*

References

[1]Wescott, C. C., *pH Measurement*, Academic Press, 1978, pp. 27-30.

Unit 12: pH Control System Troubleshooting

UNIT 12

pH Control System Troubleshooting

This unit describes a general method for logically eliminating potential causes of a system problem.

Learning Objectives—When you have completed this unit you should:

A. Know how to efficiently locate the source of the problem.

B. Be aware of hardware requirements for troubleshooting.

12-1. Measurement and Reagent System

An efficient trouble-shooting technique should systematically eliminate portions of the measurement installation depicted in the functional block diagram of Fig. 12-1. The electrodes are the most frequent, the electrode wiring is the second most frequent, and the pH meter and transmitter are the least frequent sources of pH measurement problems. Normally, troubleshooting should begin at the most likely source. However, the pH meter and transmitter are so easily checked that often this is done first. A pH electrode simulator, which consists of a high-impedance millivolt source, can be connected in place of the electrode wires and used for a detailed performance check. The integrity of the meter and transmitter can be verified without the simulator by switching the meter to standby. The meter zero and standardization adjustment should each be able to change the reading by more than one pH unit each side of 7 pH. If the pH reading is set to 6 pH by the standardization for a 25° setting of the manual temperature compensator dial, a 0° setting should read 5.9 pH and a 100° setting should read 6.2 pH. If the meter span or slope adjustment is reduced to 80%, the meter reading should change from 6 to 5.8 pH (Ref. 1). The control room indicator should be checked at the same time to verify the integrity of the transmitter, its wiring, and the control room indicator.

The field electrodes should be checked at the electrode assembly to test the electrodes, the preamp, and the wiring to and from the preamp. If buffer solutions are used, a spare set of electrodes can be used to individually replace, and therefore locate, a faulty

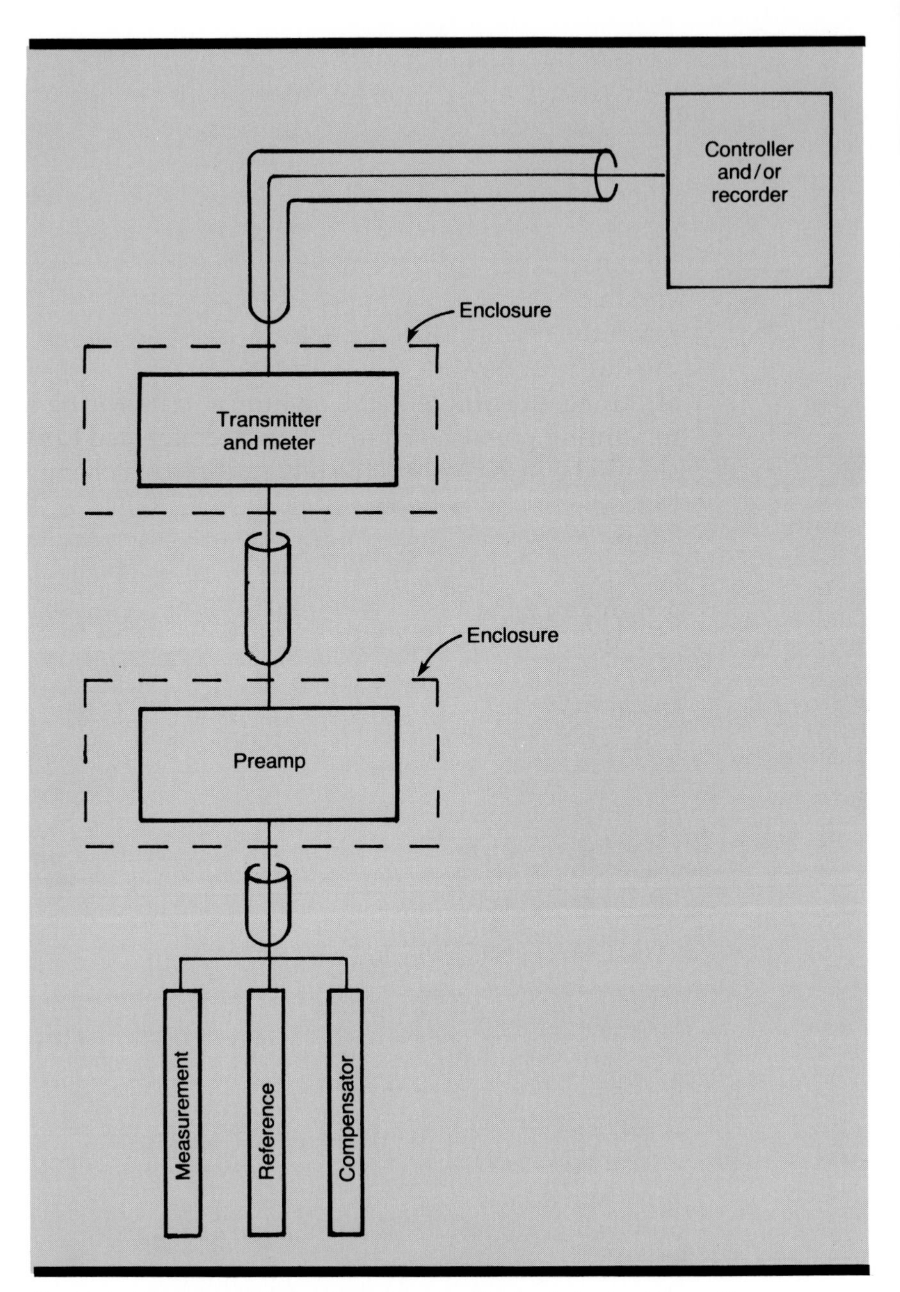

Fig. 12-1. Efficient troubleshooting will systematically eliminate portions of the measurement installation as the source of the problem.

electrode. The symptoms of the measurement problem can be used in conjunction with Table 4-4 to diagnose the nature of the electrode problem so that preventive measures can be taken. If a sample of unknown pH is used, the spare set of electrodes can be used to verify whether there is an electrode problem by noting whether the pH reading is the same for each of the four combinations of the used and spare electrodes. Any disagreement can be used to logically eliminate the measurement electrodes or reference electrodes as the source of the problem. However, the method cannot tell whether the used or spare electrode is at fault since the pH of the sample is unknown (Ref. 2). The speed of the response of the electrodes should be checked to verify that the hydrated layer of the measurement electrode is in good shape and that electrode coating has not started. The electrodes should reach the final pH reading within 10 seconds after placement in the buffer or sample. If replacement of each electrode does not eliminate the problem, the preamp and its wiring is the only remaining potential source. A wire connected to earth ground should be inserted into the buffer. If the pH changes, a ground loop exists somewhere in the preamp and its wiring system. A common mistake is to ground the shield because this is a common practice for instrument wiring systems. This creates a ground loop because the shield is connected to the instrument circuit ground. Sometimes a solution ground wire is installed to make sure ground current bypasses the higher resistance path through the reference electrode back to the instrument circuit ground. The resistance R10 in Fig. 4-4 is small compared to the sum of the resistances R7, R6, R5, and R4. A broken or missing solution ground wire will cause a ground current through the reference electrode.

The integrity of the preamp, its enclosure, and its wiring can be selectively tested by replacing each item with a spare that has been verified as satisfactory on another installation. The enclosure must also be replaced because it may be the source of an unwanted earth ground. This particular problem occurred in two separate installations when the preamp fiberglass enclosure was replaced with an aluminum enclosure that provided greater terminal access area for maintenance. The erratic reading that resulted did not disappear when the electrodes, preamp, meter and transmitter, and all wiring were replaced.

An excessive reagent delivery delay can be determined by making an open loop test. The total loop dead time is the time from the movement of the manual output of the pH controller to the first noticeable change in pH measurement. If the measurement is noisy, the manual output change must be large enough to cause a pH measurement change larger than the noise band. A CRT trend plot of a distributed control system cannot be used because the dead time is usually smaller than the reporting time of the data highway. A variable speed recorder is needed. The turnover time from the agitator specification sheet and a guess at the electrode time constant from Table 4-5 is summed and subtracted from the total loop dead time to give the reagent delivery delay. The source of most of the delay can be blamed on backfilling of the reagent line if the delay is much greater when the reagent control valve opens after being fully closed for a while. The control valve stroke can be monitored to make sure a restriction in the pneumatic accessories is not causing a slow stroke, but the more likely additional source of reagent delivery delay is partial flow in the pipeline between the valve and injection point.

A leaking reagent valve is difficult to detect if a reagent flow meter is not installed. The influent flow has to be stopped and the pH recorded to see if it continues to drop for a leaking acid valve or continues to rise for a leaking base valve. For a dual acid and base reagent system, the acid valve is leaking if the pH controller keeps the base valve open and vice versa. An overlap or too large of a gap in the split range of dual reagents can be diagnosed by comparing a recording of the pH with Figs. 6-2a and 6-2b.

12-2. Control System

A tuning problem can be diagnosed by looking at a trend recording of the dynamic closed loop response. If a nonlinear controller is used, a limit cycle will result if the breakpoints are set too close together or the notch gain is too large and a slow drift will result if the breakpoints are too far apart or the notch gain is too small. If the oscillation period is much larger than the ultimate period, the controller has too much reset action and has gone into a reset cycle. If the oscillation period is smaller than the ultimate period or turns back before reaching setpoint, the controller has too much rate action. If the oscillation period is much larger than the ultimate period estimated from the turnover time, the electrode response is too slow due to coating or dehydration, the reagent delivery delay is too large due to backfilling or partial flow, or a dead zone exists in the vessel due to an incomplete axial flow pattern. If the measurement stays too long out on the flat portion of the titration curve for a large disturbance, the proportional band is too large.

If the controller is properly tuned, the source of a persistent pH oscillation could be reagent pressure fluctuations, reagent control valve hysteresis, pH measurement noise, or influent disturbances. If the oscillation goes away when the reagent control valve is placed in manual, control valve dither due to hysteresis caused the oscillation. If the oscillation goes away when the reagent control valve is closed, reagent pressure fluctuations caused the oscillation. If the reagent valve is left open on manual and the influent flow is stopped, and the oscillation persists, pH measurement noise caused the oscillations. If the measurement and reagent systems did not cause the oscillations, influent disturbances are to blame by default.

Exercises

12-1. *How can a pH meter be functionally tested without electrodes or an electrode simulator?*

12-2. *Why should the electrode response time be noted when testing the electrodes?*

12-3. *What are the causes of a slow loop oscillation?*

12-4. *What are the causes of pH measurement noise?*

References

[1]Wescott, C. C., *pH Measurement*, Academic Press, 1978, pp. 140-141.
[2]*Ibid.*, pp. 143-145.

Appendix A: Suggested Readings and Study Materials

APPENDIX A

Suggested Readings and Study Materials

Independent Learning Modules

Murrill, P. W., *Fundamentals of Process Control Theory*. Instrument Society of America, 1981.

Handbooks

Liptak, B. G., *Instrument Engineers Handbook: Vol. 1—Process Measurement*, Chilton Book Co., Second Ed., 1983.

Textbooks

Bates, R. G., *Determination of pH Theory and Practice*, John Wiley & Sons, 1964.

McMillan, G. K., *Tuning and Control Loop Performance*, Instrument Society of America, 1983.

Shinskey, F. G., *pH and pION Control in Process and Waste Streams*, John Wiley & Sons, 1973.

Skoog, D. A. and West, D. M., *Principles of Instrumental Analysis*, Saunders College, Second Edition, 1980.

Wescott, C. C., *pH Measurement*, Academic Press, 1978.

Papers

Gray, D. M., "Microprocessor Characterizes pH Ahead of Controller for Easy Tuning," *Control Engineering*, Jan. 1980, pp. 79-81.

McMillan, G. K. "A Survey of pH Control Problems and Solutions—A Magical Mystery Tour", *In Tech*, Sept. 1984.

Trevathan, V. L., "Advanced Control of pH," *ISA Advances in Instrumentation*, Vol. 33, Part 2, 1978, pp. 69-82.

Appendix B: Solutions to All Exercises

APPENDIX B

Solutions to All Exercises

UNIT 1

1-1. The hydrogen ion concentration must decrease by ten to the fifth power. Since the hydrogen ion concentration is proportional to the acid flow for a strong acid and base system, the acid flow must also decrease by the same factor.

1-2. The slope of the titration curve, and hence the pH process gain, is less at 1 than at 6 pH, so that changes in net acid or base concentration along the abscissa due to control valve hysteresis translate into smaller pH changes.

1-3. The pH process gain increases by a factor of ten to the fifth power since the difference between 1 and 6 pH is 5 pH and the pH gain increases by a power of ten for each pH unit increase towards 7 pH.

1-4. The titration curve slope and hence the pH process gain is greatest at 7 pH for a strong acid and base.

UNIT 2

2-1. The influent density is 1100 gm per liter for a 1.1 specific gravity. The molecular weight is 40.01 per Table 2-3b. The valence or charge of sodium hydroxide is 1 per Table 2-2. Equation (2-3a) is solved for the weight fraction of sodium hydroxide in the influent and multiplied by 100%.

$$x_{ii} = \frac{M*N}{z*d} = \frac{40.01*2.75}{1*1100} = 0.10$$

weight percent = 0.10 * 100% = 10%

2-2. The hydrochloric acid concentration equals the sodium hydroxide concentration in the effluent for complete neutralization since they are both fully ionized. The effluent density is 1080 gm per liter for a 1.08 specific gravity. The molecular weight of hydrochloric acid is 36.46 and the molecular weight of sodium hydroxide is 40.01 per Tables 2-3a&b. The valence or charge of hydrochloric acid and sodium hydroxide is 1 per Table 2-1. Equation (2-3a) is ratioed for the weight fraction of hydrochloric acid to sodium hydroxide in the effluent. Equation (2-4c) is used to calculate the required reagent flow in pph.

Since $N_r = N_i$ and $Z_r = Z_i$

$$\frac{X_{re}}{X_{ie}} = \frac{\dfrac{M_r*N_r}{Z_r*d}}{\dfrac{M_i*N_i}{2_i*d}} = \frac{M_r}{M_i} = \frac{36.46}{40.01} = 0.9125$$

$$F_r = \frac{x_{re}}{x_{ie}} * \frac{x_{ii}}{x_{ee}} * F_i$$

$$F_r = 0.9125 * \frac{0.10}{0.20} * 100 = 45.6\text{pph}$$

2-3. At 7 pH and for a pkw equal to 14, the concentration of hydrogen ions equals the concentration of hydroxyl ions. The concentration of acid ions must then equal the concentration of base ions. Since a strong acid or base is completely ionized, the concentration of the acid or base must be equal if the solution temperature is 25°C.

2-4. Since a weak acid or base is partially ionized, the concentration of the acid only equals the concentration of the base if the acid dissociation constants are equal to the base dissociation constants in addition to the solution temperature being 25°C.

2-5. At 50°C, the pKw is 13.26 per Table 2-4. Equation (2-16) shows that for equal hydrogen and hydroxyl ion concentrations, the hydrogen ion concentration is equal to the square root of the ionic product or the pH is equal to half of the pKw.

$$pH = \frac{pK_w}{2} = \frac{13.26}{2} = 6.63$$

UNIT 3

3-1. The systems ranked from least to greatest slope and hence reagent sensitivity and pH process gain at 6 pH are:

(1) weak acid and weak base

(2) weak acid and strong base

(3) strong acid and weak base

(4) strong acid and strong base

3-2. The best pH setpoint would be at a pH equal to the pKa since this is the midpoint of the flat portion of the titration curve.

3-3. The caustic solution would absorb carbon dioxide which would add a buffering effect that may not be in the process, the number of data points may be insufficient near the equivalence point due to the steep slope, and the error at the high pH end of the scale may be different than the field due to sodium ion error.

3-4. The interhalving-search method will always converge since there is only one zero crossing of the charge balance if the pH is within the original search interval selected and if the arithmetic precision of the computer is sufficient for the pH system.

3-5. The laboratory titration curve would be more accurate due to the uncertainty as to the activity coefficients from high ion interaction.

3-6. The lime would not be completely dissolved with its ions into solution in the few seconds it takes the reagent to reach the electrodes in the pipeline.

UNIT 4

4-1. The internal fill is at 6 pH per Eq. (2-6), so that 6 instead of 7 pH should be used in Eq. (4-4), and the isopotential point is shifted from 7 to 6 pH.

4-2. The slope is the absolute temperature term in Eq. (4-4) that is multiplied by the deviation of the pH from the isopotential point.

$$\text{slope} = 0.1984*(T + 273.16) = 0.1984*(100 + 273.16) = 74.04 \text{ mv/pH}$$

4-3. An erratic reading can be caused by measurement bulb abrasion, dehydration, or etching, pure water, or gas bubbles.

4-4. A slow response can be caused by measurement bulb abrasion, dehydration, etching, or coating, nonaqueous solutions, or low sample velocity.

4-5. The akalinity error causes the shortened span or bending over of the millivolt versus pH line at the high end of the scale.

4-6. A constant pH reading is most likely due to a broken measurement electrode.

4-7. The measurement electrode terminal could be shorted to ground or to the reference electrode terminal, the pH meter switch could be in the standby position, or the circuit could be open due to a broken electrode wire or a complete nonconductive coating of either electrode.

4-8. Since no known electrode failure could cause such an extreme negative millivoltage, the source must be either a nonaqueous acid solution or an open circuit in the transmitter input or output wiring.

4-9. The transportation delay is calculated via Eq. (4-10) after calculating the cross sectional area as 0.0233 feet and converting the flow from 100 gpm to 13.37 cu ft per minute.

$$TD_p = \frac{A*L}{F} = \frac{0.0233*100}{13.57} = 0.17 \text{ minutes} = 10 \text{ seconds}$$

4-10. The methods of reducing electrode coating problems are ultrasonic, mechanical brushing, jet washing, and flushing by a high sample velocity.

UNIT 5

5-1. The agitator pumping rate is 1875 gpm, the equipment dead time is 0.5 minutes, and the equipment time constant is 9.5 minutes.

$$N_q = \frac{0.4}{\left(\frac{D_i}{D_t}\right)^{0.55}} = \frac{0.4}{\left(\frac{2}{4}\right)^{0.55}} = 0.586$$

$$Q_a = N_q * N * D_i^3 = 0.586 * 400 * 8 = 1875 \text{ gpm}$$

$$TD_p = \frac{V}{Q_a + Q_t} = \frac{1000}{1875 + 100} = 0.5 \text{ minutes}$$

$$TC_p = \frac{V}{Q_t} - TD_p = \frac{1000}{100} - 0.5 \text{ minutes} = 9.5 \text{ minutes}$$

5-2. The unagitated vertical tank provides the best filtering action since the equipment dead time is 20 to 30% of the residence time, which means the filter time constant is 70 to 80% of the residence time.

5-3. The static mixer has the smallest magnitude of equipment dead time since it has the smallest volume.

5-4. The pond or lagoon has the largest magnitude of equipment dead time since it has the largest volume.

UNIT 6

6-1. Equation (6-1c) is used to calculate the allowable reagent error. The maximum control valve hysteresis must be less than this error per Eq. (6-2).

for Fig. 6-1a:

$$E_v < 80 * \frac{B}{A} = 80 * \frac{0.1}{0.8} = 10\%$$

for Figure 6-2a:

$$E_v < 80 * \frac{B}{A} = 80 * \frac{0.01}{0.8} = 1\%$$

6-2. The reagent error is doubled and the pH error is increased by a factor of 20. pH control loops require special control valves.

6-3. If the leak is caused by a stuck stem, a valve position indicator in the control room could be used but valve position transmitters and indicators are rarely installed. If the leak is caused by valve seat corrosion or erosion, the leak is very difficult to detect and can result in significant waste of both reagents. A monitoring of the pH controller output would show a larger than normal base reagent demand. If the influent flow is stopped and the setpoint is above the acid reagent pH, the base reagent valve will remain open. An acid and base reagent system should have flowmeters to detect leaks.

6-4. Equations(6-4a) and (6-4b) are used to calculate that the signal portions are 80 and 20% for the acid and base valves respectively. One of the valve actions must be reversed since the reagents have opposite effects on the pH measurement. If the acid valve is reversed, the resulting split range works as shown in the table below, which shows a split range gap of 2% at the transition point.

$$S_1 = \frac{F_{1max} * N_1}{F_{2max} * N_2} * S_2 = \frac{100 * 10}{100 * 2.5} * S_2 = 4 * S_2$$

$$S_1 + S_2 = 100$$

$$4 * S_2 + S_2 = 100$$

thus: $S_2 = 20$ and $S_1 = 80$

Output (%)	Acid Valve	Base Valve
0	Wide open	Closed
39.5	Half open	Closed
79	Closed	Closed
81	Closed	Closed
90.5	Closed	Half open
100	Closed	Wide open

6-5. The prestroke dead time due to hysteresis is approximately equal to the valve hysteresis divided by the signal rate of change. For 10% valve hysteresis and a 1% per second signal rate of change, the prestroke dead time is about 10 seconds.

6-6. Equation (6-6) is solved for the pulse width and used with an allowable reagent error of 0.01 per influent pph from Fig. 6-1b.

$$A_f = 0.01 * 200 = 2 \text{ pph}$$

$$TC_p = \frac{V}{Q} - TD_p = 10 - 0.5 = 9.5$$

$$T_p = \frac{TC_p * \pi * A_f}{A_p} = \frac{9.5 * \pi * 2}{100} = 0.6 \text{ minutes}$$

UNIT 7

7-1. The advantages of a properly installed caustic reagent dilution system are:

(a) Larger control trim size that is less likely to plug.

(b) Smaller reagent transportation time delay

(c) Greater injection velocity

(d) Greater injection Reynolds number

(e) Lower freezing point

(f) Lower corrosion rate for steels

7-2. A high buffer capacity means that wash water or contaminant coatings on the electrodes would not change the pH of the test solution.

7-3. From Fig. 7-1, the residence time must be 20 times the dissolution time or 640 minutes for the dry lime in Fig. 7-2.

7-4. The reagent could be diluted to increase the volumetric reagent flow rate or the reagent pipe could be purged with a gas to prevent backfilling.

UNIT 8

8-1. The coating on the electrodes will increase the electrode time constant.

Since a fraction of this time constant becomes loop dead time and the peak error is reached in one dead time, the time to the first peak is increased.

8-2. At best, the nonlinear controller gain matches the pH process gain at three points on the titration curve. The nonlinear controller has three points on the titration curve. The nonlinear controller has three additional settings (lower and upper breakpoints and notch gain) that may require some field adjustment due to the inadequacy of the straight line approximation. The linear reagent demand controller eliminates the nonlinear effects of the titration curve and restores the full potential of the controller to that dictated by the loop dead time to time constant ratio. The linear reagent demand controller is tuned like any other conventional linear controller.

8-3. If the correction for an influent pH disturbance must start before a time equal to one loop dead time has transpired, pH feedforward control should be added.

8-4. The acid error below 0 pH corresponds to a large reagent error due to flatness of the titration curve. The error gets worse with time so that compensation for the error is not possible. The pH measurement will be higher than the actual pH so that the feedforward signal will be less than required. While the error is in the safe direction, the performance of pH feedforward control will probably not be greater than that obtained by flow feedforward that is already incorporated into a linear reagent demand controller. The additional cost of pH feedforward controller is probably not justified.

8-5. The reagent should be added to the circulation pump suction to form a fast inline feedback loop. The circulation flow is normally held constant and the pH disturbance from the sump is normally slow compared to the inline loop dead time due to the equipment time constant the sump volume. Reagent addition to the sump would add the dead time of the unagitated sump to the pH control loop. Reagent addition to the top of the column would add the transportation delay from the top to the sump and the equipment dead time of the sump to the pH control loop.

8-6. The adaptive algorithm is not forced to continually compensate for an operating point nonlinearity and can be designed to concentrate on a time varying nonlinearity. If the titration curve is fixed, the linear reagent demand controller compensates for the operating point nonlinearity. The only justification for an adaptive controller is to compensate for a time varying nonlinearity caused by a changing titration curve.

UNIT 9

9-1. The equipment time constant is the residence time minus the turnover time for continuous control per Eq. (5-2) and is the mixing time divided by 5 minus the turnover time for batch control per Eq. (5-3a). The ratio of the small (electrode) time constant to the large (equipment) time constant is 0.01 for continuous and 0.1 for batch control. The effective dead time factor for the electrode time constant is about 0.9 for continuous and 0.65 for batch control per Fig. 9-1. The equivalent dead time of 0.09 minute for continuous and 0.065 minute for batch control is summed with the equipment dead time to give the total loop dead time of 0.59 minute for continuous and 0.565 minute for batch control. Equations

(9-1) and (9-2) are used to calculate the ultimate periods for the loop dead time and largest time constant as follows:

For continuous control:

$$T_u = 2 * \left[1 + \left[\frac{TC}{TC + TD}\right]^{0.65}\right] * TD = 2 * \left[1 + \left[\frac{19}{19 + 0.59}\right]^{0.65}\right] * 0.59 = 2.34 \text{ minutes}$$

For batch control:

$$T_u = 2 * \left[1 + \left(\frac{TC}{TD}\right)^{0.65}\right] * TD = 2 * \left[1 + \left(\frac{1}{0.57}\right)^{0.65}\right] * 0.57 = 2.78 \text{ minutes}$$

9-2. The product of the open loop gains are calculated by Eq. (9-4i). Note that the valve and measurement gains calculations can use a full scale change in signal because they are linear but the pH process gain calculation must be made for a change within the control band because it is nonlinear. The pH process gain is equal to the control band divided by the allowable reagent error band. Equation (9-3a) is used to calculate the ultimate proportional band and Eqs. (9-5d) to (9-5f) are used to calculate the mode settings. Note that the reset setting is the inverse of the integral time.

$$K_v * K_x * K_p * K_m = \frac{\Delta F_r}{\Delta O_c} * \frac{(\Delta F_r/F_i)}{\Delta F_r} * \frac{\Delta pH}{(\Delta F_r/F_i)} * \frac{\Delta I_c}{\Delta pH}$$

$$K_v * K_x * K_p * K_m = \frac{400 \text{ pph}}{100\%} * \frac{1}{200 \text{ pph}} * \frac{1 \text{ pH}}{0.01} * \frac{100\%}{10 \text{ pH}} = 20$$

$$PB_u = \frac{100 * T_u * K_v * K_x * K_p * K_m}{2\pi * TC} = \frac{100 * 2.32 * 20}{2\pi * 19.5} = 38\%$$

$PB = 1.7 * PB_u = 1.7 * 38 = 65\%$

$T_i = 0.5 * T_u = 0.5 * 2.32 = 1.16$ min/repeat

reset $= 1/T_i = 0.86$ repeats/min

$T_d = 0.1 * T_u = 0.1 * 2.32 = 0.23$ min (rate)

UNIT 10

10-1. Equation (4-4) is used to calculate the millivolt potential developed at 25°C and 40°C. The difference between these potentials is the millivolt error which can be converted to a PH error by the same equation. A temperature compensator is not needed since the 0.1 pH error due to temperature is small compared to the control band. If the effluent temperature stays near 40°, the manual temperature compensator can remove most of this error.

$E_1 - E_2 = 0.1984*(T + 273.16)*(7 - pH_1)$

at 25° and 5 pH:

$E_1 - E_2 = 0.1984*(25 + 273.16)*(7 - 5) = 118.31$ mv

at 40° and 5 pH:

$E_1 - E_2 = 0.1984*(40 + 273.16)*(7 - 5) = 124.26$ mv

$E_{mv} = 124.26 - 118.31 = 5.95$ mv

$E_{mv} = 0.1984*(25 + 273.16)*(E_{pH})$

$5.95 = 59.16 * E_{pH}$

$E_{pH} = 0.1$ pH

10-2. An injector type of electrode cannot be installed in a gravity overflow pipeline because the pipeline will not always flow full. The velocity in a flow chamber typically cannot be made great enough for gravity flow to reduce the rate of coating without making the chamber cross sectional area so small that plugging becomes a problem. The best solution is a submersion electrode assembly with a jet washer.

10-3. Pulse control of the valve is normally less expensive than dilution for increasing the flow area in the valve to reduce plugging. For this case, the savings is greater because dilute sulfuric acid will require special materials of construction.

10-4. The alternatives are two well-mixed tanks in series with individual pH feedback loops, one pH feedforward loop and one well-mixed tank with a pH feedback loop, or one well-mixed tank with a linear reagent demand feedback loop.

10-5. The simplified Eq. (10-1d) can be used for this well-mixed tank. The peak reagent error is equal to the open loop reagent for each vessel multiplied by 0.026. The reagent error band per pph of influent flow for the 1 pH control band is 0.01 from Fig. 6-1b. The allowable peak reagent error is equal to one half of the reagent error band multiplied by the influent flow. The peak error for the first vessel is small enough to keep the pH within the control band. However a linear reagent demand controller must be used to achieve this level of performance. Two vessels would be required if pH controllers were used because the margin of error for nonlinearity is too small.

$E_p = 0.5 * B * F_i = 0.5 * 0.01 * 200 = 2$ pph

for vessel number 1:

$$E_p = \frac{TD}{TC} * E_o = \frac{0.5}{19.5} * 50 = 1.3 \text{ pph}$$

for vessel number 2:

$$E_p = \frac{TD}{TC} * E_o = \frac{0.5}{19.5} * 1.3 = 0.04 \text{ pph}$$

UNIT 11

11-1. The alkalinity error will cause a nonlinear electrode response that can be better matched by the meter curve if the isopotential point is set at 8 pH.

11-2. The process sample pH depends upon another pH measurement for identification, the pH may change due to absorption of carbon dioxide or contamination, the pH may be more sensitive to temperature, and the sample may be hazardous.

11-3. The sample may start to boil and cause erratic readings due to bubbles, process fluid degradation, and electrode high-temperature failure.

11-4. The injector pH probe can be installed in a piping elbow or tee as long as the electrode tips are not pointing up.

11-5. Welding rods, bolts, and even wrenches may be in the pipelines or vessels where the electrodes are installed.

11-6. pH feedforward control is corrected by a summer whereas flow feedforward control is corrected by a multiplier. A pH controller can override a flow feedforward signal by multiplication by zero.

UNIT 12

12-1. The meter is placed in standby and the calibrations adjustments are moved to check their functionality. The reading must be moved from the isopotential point by the standardization adjustment to check the operation of the span and temperature compensation adjustments. The meter zero cannot be used for this purpose because it shifts the isopotential point to the new reading.

12-2. A slow electrode response is symptomatic of coatings or damage to the hydrated gel layer of the pH electrode. It can lead to slow loop oscillations, large integrated errors, and eventual total loss of the electrode response.

12-3. A slow loop oscillation can be caused by too much reset action, a slow electrode response, a large reagent delivery delay, or a vessel dead zone.

12-4. pH measurement noise can be caused by incomplete backmixing, gas bubbles, measurement bulb abrasion, dehydration, or abrasion, and electrical interference.

Appendix C: Glossary of pH Control System Terminology

APPENDIX C

Glossary of pH Control System Terminology

activity—Ratio of escaping tendency of the component in solution to that at a standard state. The ion concentration multiplied by an activity coefficient is equal to the ion activity.

association—The combining of ions into larger ion clusters in concentrated solutions.

asymmetry potential—The difference in potential between the inside and outside pH sensitive glass layers when they are both in contact with 7 pH solutions. It is caused by deterioration of the pH sensitive glass layers or contamination of the internal fill of the measurement electrode.

dissociation—The breaking apart of a molecule into its component ions in a solution.

diffusion—The movement of ions from a point of high concentration to low concentration.

equivalence point—Point on the titration curve where the acid ion concentration equals the base ion concentration.

ionic strength—Effective strength of all ions in a solution that is equal to the sum of one half of the product of the individual ion concentration and their ion valence or charge squared for dilute solutions.

isopotential point—Point on the millivolt versus pH plot at which a change in temperature has no effect. It is at 7 pH and zero millivolts unless shifted by the standardization and meter zero adjustments or an electrode assymetry potential.

migration—The movement of ions from an area of the same charge to an area of opposite charge.

molal units—Concentration units defined as the number of gm-moles per 1000 gm of solvent.

molar units—Concentration units defined as the number of gm-moles of the component per liter of solution.

moles—Number of molecular weights which is the weight of the component divided by its molecular weight.

neutral point—Point on the titration curve where the hydrogen ion concentration equals the hydroxyl ion concentration.

normality—Concentration units defined as the number of gram-ions of replaceable hydrogen or hydroxyl groups per liter of solution. A shorter notation of gram-equivalents per liter is frequently used.

titration curve—A plot with pH as the ordinate and units of reagent added per unit of sample as the abscissa.

Appendix D: Review of Algebra with Logarithms

APPENDIX D

Review of Algebra with Logarithms

$\log(m*n) = \log(m) + \log(n)$

$\log\left(\frac{m}{n}\right) = \log(m) - \log(n)$

$\log(10^p) = p*\log(10) = p$

$\text{antilog}(p) = 10^p$

$\log(m*10^p) = \log(m) + p$

Therefore, to convert from hydrogen activity to pH: (The same procedure is used to convert from an acid and base dissociation constant to a pKa and pKb, respectively).

$pH = -\log(a_H)$

$pH = -\log(m*10^{-p})$

$pH = -(\log(m) - p)$

For example, if the hydrogen activity is 0.0003:

$pH = -\log(3*10^{-4})$

$pH = -(\log(3) - 4) = 3.523$

And to convert back from pH to hydrogen activity:

$a_H = -\text{antilog}(pH)$

$a_H = -\text{antilog}(-(\log(m) - p))$

$a_H = m*10^{-p}$

For example, if the pH is 3.523:

$a_H = -\text{antilog}(-(0.477 - 4))$

$a_H = -\text{antilog}(-(\log(3) - 4))$

$a_H = 3*10^{-4}$

Appendix E: FORTRAN Subroutine to Simulate a pH Process

APPENDIX E
FORTRAN Subroutine to Simulate a pH process

```
      SUBROUTINE ZPH3(Y, C, P, W, M, H, PK1, PK2, PK3, N)

      REAL C(N), P(6),W(N),M(N),H(N),PK1(N), PK2(N), PK3(N)

C

C     CALCULATED OUTPUTS:
C     Y IS THE PH WITH OPTIONAL ELECTRODE ERROR
C     C IS THE CONCENTRATION IN NORMALITY UNITS
C     DATA INPUTS:
C     P(1) IS THE DISSOCIATION CONSTANT FOR WATER
C     P(2) IS THE ELECTRODE ERROR INDICATOR
C     P(3) IS THE AVERAGE DENSITY
C     P(4) IS THE LOWER PH LIMIT
C     P(5) IS THE UPPER PH LIMIT
C     P(6) IS THE PH SEARCH ERROR
C     W IS THE WEIGHT PERCENT
C     M IS THE MOLECULAR WEIGHT
C     H IS THE NUMBER OF HYDROGEN OR HYDROXL IONS
C     PK1 IS THE FIRST DISSOCIATION CONSTANT
C     PK2 IS THE SECOND DISSOCIATION CONSTANT
C     PK3 IS THE THIRD DISSOCIATION CONSTANT
C     N IS THE TOTAL NUMBER OF ACIDS AND BASES
C
C     CALCULATE THE CONCENTRATION OF EACH ACID AND BASE
C
      PKW = P(1)
      DO 10 J = 1,N
      C(J) = 1000.*H(J)*P(3)*W(J))/M(J)
   10 CONTINUE
C
C     ASSIGN THE UPPER AND LOWER LIMITS OF PH SEARCH
C
      PHL = P(4)
      PHH = P(5)
C
D     CALCULATE THE MID POINT PH AND SEARCH ERROR
C
   20 PHM = (PHL + PHH)/2.
      E = (PHH - PHL)/2.
      I = 1
C
C     CALCULATE THE HYDROGEN AND HYDROXYL ION
      CONCENTRATIONS
C
      QE = 10.**(-PHM) - 10.**(PHM - PKW)
C
C     SET PARAMETER S SIGN AS INDICATOR OF ACID OR BASE
C
      DO 30 J = 1,N
      S = SIGN(1.,H(J))
C
C     CALCULATE EFFECT OF FIRST DISSOCIATION CONSTANT
C
```

```
      P1 = 10.**(S*(PHM - PK1(J)))
C
C     CALCULATE EFFECT OF SECOND AND THIRD DISSOCIATION
      CONSTANT
C
      IF (ABS(H(J)).GT.1.) P2 = 10.**(S*(PHM - PK2(J)))
      IF (ABS(H(J)).GT.2.) P3 = 10.**(S*(PHM - PK3(J)))
C
C     CALCULATE THE ACID AND BASE ION CONCENTRATIONS
C
      QI = C(J)/(1. + P1)
      IF (ABS(H(J)).GT.1.)  QI = C(J)*(1. + 0.5*P2)/(1. + P2*(1. + P1))
      IF (ABS(H(J)).GT.2.)  QI = C(J)*(1. + 0.33*P3*(2. + P2))
      $                         /(1. + P3*(1. + P2*(1. + P1)))
C
C     CALCULATE THE NET CHARGE BALANCE
C
      QE = QE + QI
   30 CONTINUE
C
C     REDUCE THE SEARCH INTERVAL BY ONE HALF
C
      IF (QE.GT.0.) I = -1
      IF (I.LT.0) PHL = PHM
      IF (I.GT.0) PHH = PHM
C
C     CHECK ERROR TO REPEAT OR END PH SEARCH
C
      IF (E.LT.P(6)) GO TO 40
      GO TO 20
C
C     CORRECT FOR LOW AND HIGH PH ELECTRODE ERROR
C
   40 PHE = 0.0
      IF (PHM.GT.1.0) GO TO 50
      IF (P(2).EQ.1.0.OR.P(2).EQ.3.0) PHE = -0.049999 + 0.135999*PHM
      $        + 0.002000*PHM**2. - 0.01600000*PHM**3. -
      0.071999*PHM**4.
   50 IF (PHM.LT.8.0) GO TO 60
      IF (P(2).EQ.2.0.OR.P(2).EQ.3.0) PHE = 50.64536 - 21.20648*PHM
      $        +3.316879*PHM**2. - 0.2305096*PHM**3. +
      0.006042*PHM**4.
   60 Y = PHM - PHE
      RETURN
      END
```

Index

John C. Pfeiffer, P.E.
Pfeiffer Engineering Co.
560 Garden Drive
Louisville, Kentucky 40206

INDEX